# Appendices to the Atmos Book of Pipeline Simulation

Moemen Metwally, AMIChemE

*Application Support Engineer, Atmos International*

Dr. Jason Modisette

*Chief Scientist, Atmos International*

*Editors*

Andy Hoffman

Dr. Jun Zhang

First published 2022



*1st edition*

Authors Moemen Metwally and Dr Jason Modisette

Edited by Andy Hoffman and Dr Jun Zhang

Cover design by Sam Edge

ISBN 978-1-7397829-1-7

# TABLE OF CONTENTS

# OVERVIEW OF THE APPENDICES

These appendices were compiled into a separate book to cover a number of topics that may have distracted the reader if included in the main text of the Atmos Book of Pipeline Simulation yet are of great interest and worth turning our attention to as well.

In these appendices, we dive deeper to look at what today's energy transitions mean for us today (Appendix A) and consider the role of pipelining and hydraulic engineering in the grand sweep of human history (Appendix B).

Then, we present a primer on the core principles of how fluids flow and behave in pipelines (Appendix C) and run through some important fluids transported in pipelines (Appendix D). We devote particular attention to hydrogen as an emerging energy carrier in places, and challenges to overcome in deploying it, striving to contextualize the debate around its role in the future (Appendix E).

Finally, we challenge the reader with a quiz (Appendix F).

# APPENDIX A – THE ENERGY TRANSITION

*"We cannot solve our problems with the same thinking we used when we created them."* – Albert Einstein

## A1 – TO BE, OR NOT TO BE SUSTAINABLE

Do pipelines do good in the world? In a time when building a pipeline has been politicized, the pipelining industry has an image problem in environmentally conscious circles, becoming one of several elephants in the room that must be addressed. To understand how pipeline infrastructure works, and renew our commitment to averting its possible harms, we must ask why it's there at all. How do hydrocarbons fit in the energy landscape of this future we all live in? Many governments now only support our sector if it works to deliver a strategic target: net zero emissions. There is pressure from investors and the public at large to respond to the challenge. Shareholders insist listed companies price carbon into any business models, demonstrating how emissions from their operations are reduced to support wider decarbonization of the economy.

**Figure 1 – Since 2000, there has been a shift towards natural gas for generating electricity**

We use energy for electricity, transport, buildings and industry. Since 2000, electricity in many places has been generated by burning natural gas rather than coal or oil. Since 2010, there has been a new transition to renewables. Transport, buildings and industry are proving tougher to decarbonize than electricity: liquid fuels still dominate transport, natural gas dominates heat. A pathway to decarbonize transport, buildings and industry is to have them use electrical power from a low-carbon grid. This approach has limits.

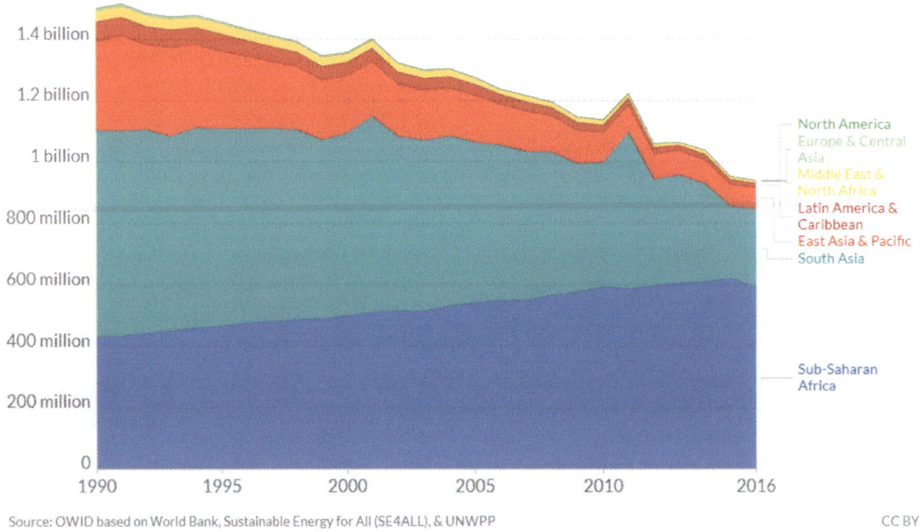


**Figure 2 – People without access to electricity (Our World in Data)**

For one, a billion people still lack access to electricity: more than one in ten. There is no sense expecting to electrify sectors where there is no electricity! As the developing world builds power infrastructure, developed countries' insatiable appetite refuses to abate, so global energy consumption is still rising. Low-carbon energy is only making a small dent in our power generation. 80% of global energy consumption is still derived from fossil fuels, and the world is relentlessly consuming more fossil fuels year-on-year.

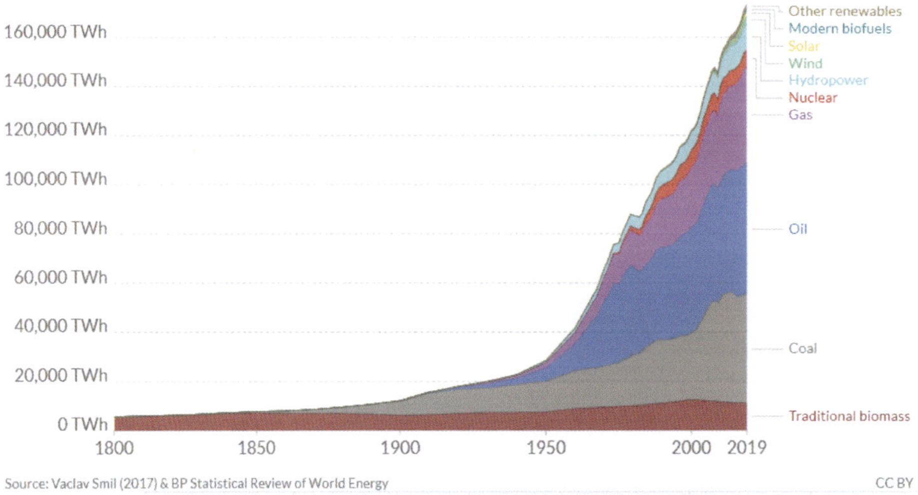


**Figure 3 – Energy consumed worldwide, by source (Our World in Data)**

Sustainability is about the long-term implications of what we do. In the context of the water industry, say, it can't be boiled down to one single question. Sustainability is as much about whether an activity makes sense at all in a particular climate, as it is about efficiency or expendable resources. In the desert, rice farming is possible by digging wells and exploiting groundwater, but it is certainly *not* sustainable. That did happen, and people eventually learnt their lesson the hard way when it depleted a non-replenishing resource: water. When water ran out, they were left with two options: force vast quantities of seawater onto paddy fields via desalination plants, or farm elsewhere.

*"Do we really need air conditioning"* seems like an uncomfortable challenge to the status quo, but our predecessors often understood how to design a building to be naturally better ventilated and cooled better than we do. Such questions are becoming the norm, if only to reduce consumption rather than eliminate it altogether. Do we really need to use so much energy? Planning an urban environment for necessary trips is simply asking *"can I buy milk without driving for miles?"* Questioning if everyone must commute at the exact same time every single day – or, for some, if they need to at all! If a neighborhood is well-designed, homes, shops and workplaces consume less.

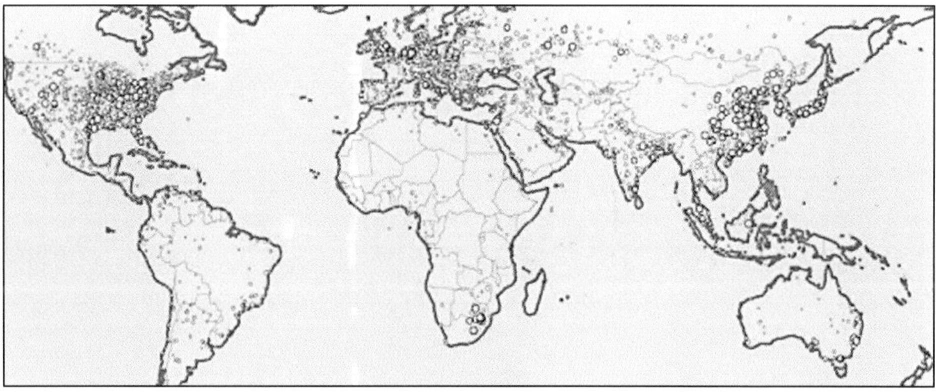

**Figure 4 – Major stationary emitters of carbon dioxide around the world. A substantial proportion of our $CO_2$ emissions come from a few thousand industrial sites (IPCC, UN)**

Relying on fossil fuels for energy in the way that society does today, particularly coal and liquids refined from oil, is broadly accepted as being unsustainable for two separate reasons. One being that the resource is expendable, sooner or later, and all the palm or corn oil in the world won't make enough biofuel to replace it for future generations. The other is that we're ruining our environment long before oil or gas run out: fossil fuels cause most of our carbon emissions.

Merely saying "*energy transition*" is enough to rile up righteous outrage in diametrically opposing circles. As a vision for the future, it is by nature loaded and speculative. We find ourselves forced to ask: *whose* energy transition? Is it to improve energy security, relying on *"our own"* resources rather than foreign imports – *"us"* being whichever country one happens to be in? Is it to wean society off fossil fuels as an emergency, shifting its reliance towards sustainable alternatives, given the alarming rate at which *"rare"* climate disasters have become the new normal? Or is it to drive human progress by very virtue of revolutionary technologies being unprecedented?

> *"Many oil and gas companies are now responding to the challenge. Their investment decisions are beginning to anticipate a world without fossil fuels beyond 2050."*
> – UK Government Whitepaper, December 2020

Incredible advances in batteries, electronics and computing dominate our collective psyche, with electric vehicles, smart grids and emerging economic trends all relying on the aforementioned insatiable appetite for electricity. Let's take a look at how people are using energy transmitted as electricity first of all, then consider transport and finally buildings and industry. Before talking about pipelines, it's vital we understand why we need them in the first place.

**Figure 5 – Distinctive yellow pipework feeds natural gas to industrial boilers**

# A2 – ENERGY FOR ELECTRICITY

People like to think that we've moved on from the 20th century, but often one doesn't need to dig too deep to find combustion in cylinders turning the gears. The internet, in practice, resides in places like warehouses on the outskirts of Dublin, whose power grid has been struggling to cope. The biggest names in tech equip their warehouses with dozens of the largest gas engines simply to provide electricity and cooling for their datacenters. Our 21st century servers hosted on *"the cloud"* are ultimately fueled by natural gas pipelines. Countries whose electric power grids are less reliant on fossil fuels have often mostly outsourced their manufacturing abroad, where more coal is burned than ever.

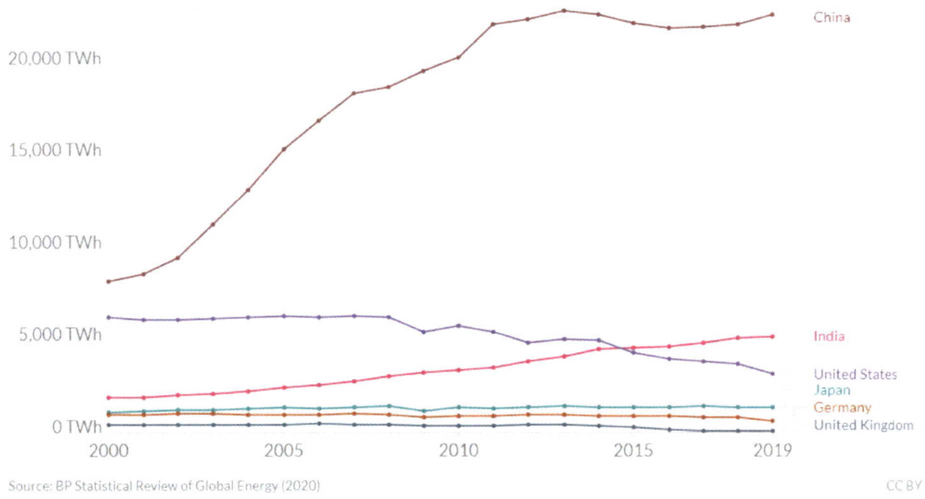


**Figure 6 – Coal consumption has been rising in India and China (Our World in Data)**

Those running the datacenters find so much heat is wasted to cool their computers from 30°C to 20°C they are left scratching their heads thinking what they can possibly do with it. They run up against thermodynamic limits on low grade heat, finding that waste is practically unavoidable. Similar theoretical limits govern how efficiently we can transport a fluid or transmit electricity.

The image we have in Britain, say, of how our electric grid works on this windy island, with the North Atlantic on one side and the North Sea on the other, is one of offshore wind. Even here where the grid has rapidly and significantly shifted its reliance towards offshore wind power, we have run up against their *intermittency*. Many other countries, like Australia, where solar power has suddenly appeared in force, are also running up against this same challenge.

**Figure 7 – The *Academik Lomonosov*, a nuclear power barge, was towed to the arctic. It is capable of generating electrical power for 200 000 people and district heating. Elena Dider / Wikimedia Commons / CC BY-SA 4.0**

The prohibitive upfront cost of nuclear power has limited its uptake, with the only country to meaningfully overcome this hurdle in recent years being China, which replaced many of its coal-fired plants and the associated particulates with latest-generation nuclear. [1] France relies on an older fleet of nuclear reactors. Advocates of nuclear power point to advances in technology that have overcome two chief obstacles: the fuel, and the waste. However, commercial and organizational impediments prevent nuclear being a silver bullet: the stubborn refusal of *small modular* to materialize keeps nuclear projects too front-loaded. Simply put, it's expensive to build, and takes too long to turn a profit. To compound matters, safety concerns following Fukushima mean even Germany is switching off its nuclear fleet permanently. Hype continues around fusion and anticipated advances that still, as ever, remain *"twenty years away"*.

Several less glamorous trends are playing out with no media frenzy. Driven by a revolution in countless technologies, and manufacturing at scale driving down costs, we are living through an age of energy transition. Unsubsidized renewables have become *cheaper* than fossil fuels for generating electricity. Mainstream figures in the finance community, like Warren Buffet, blazed the trail: institutional investors routinely fund solar and wind projects, purely on the merit of their profits. In Oman, solar power is even used for oil extraction! [2]

Even if nuclear power were built overnight to satisfy our power grid's baseload, we're still faced with a new challenge: smoothing out the spikes that come with renewables, so the power grid can handle them. In April 2019, California had to disconnect so many photovoltaic solar panels that an opportunity was missed to generate power satisfying 17 000 homes *all year*. In 2016, Gansu had to throw away over 30% of its photovoltaic and 40% of its wind power! To avoid this kind of *curtailment*, interconnection to a wider electrical grid was improved. A larger, well-interconnected power grid is inherently more reliable, and more capable of smoothing out spiky renewables.

Mainland Europe is at the heart of the world's most extensive such power grid. An undersea interconnector between Norfolk in England and Zeebrugge in Belgium is one of dozens linking the continent to itself and to its neighbors. This achieves tremendous savings in a thoroughly sensible way. On a day which is not windy enough for Denmark to feed its own grid, it can draw electrical power from its neighbors, such as Germany. Germany in turn, can avoid having to rely on its coal plants by making up for this electrical power with imports from France. France's nuclear reactors have a steady generation that can't be meaningfully ramped up to compensate, so it compensates by importing from Spain, whose solar generation might be in surplus at that time.

This arrangement is a win-win for all. It has provided confidence for investors at the periphery of the network. At one end, this power grid is expanding to include countries in North Africa, the Sahara being a good place for both solar and wind power, with long-distance power cables planned to transmit reliable power across the Mediterranean. At the other end, hydroelectric projects make sense, as parts of Scandinavia are mountainous enough to balance even their *neighbors'* grid. Soon, every member state is bound to have interconnected 10% of its electrical power generation capacity.

Eventually, this involves the small matter of rebuilding the entire backbone of our power grid, but countries like the US have such an old grid that this money may need to be spent anyway. The power grid's transmission losses, which are significant even with a *"super-grid"*, can be improved somewhat if a good regulatory framework lets neighbors trade power more locally, for an *overall* effect of transmitting it as far as needed.

Unfortunately, we find that even with good interconnection, above some percentage of renewables, energy storage inevitably becomes a limiting factor, particularly for seasonal balancing, rather than short-term frequency response. A capable well-managed *gas grid* – a pipeline network – helps us here. This is well-known to everyone working with either grid, but seemingly few outsiders.

As it stands, there is a limit to how much solar and wind power we can bring online before we start relying on fossil fuels for peaking or standby power. As matters stand today, Europe relies on natural gas – pipelined from nearby, or liquefied and tankered from afar. When someone owns a field near a city in the UK, the government pays them to line up a few gas engines sitting in wait, ready to spring into action within a few seconds or minutes. Along with factories willing to switch off their machinery at a moment's notice, this is how unsophisticated electrical power generation actually looks on the ground. And that's without even speaking about the *quality* of the power: a spinning lump of metal is much better at synchronizing a grid than clever electronics, and the manual nature of keeping the different parts of the puzzle working in tandem has already caused one major power outage across the UK.

Like any engineering solution, our state of play seems less than ideal, but it broadly works well overall. Natural gas burns cleaner than liquid and solid fossil fuels, and when used in combined cycle power plants it is exploited highly efficiently. Provided we make sure the *incredibly* potent methane doesn't leak and contribute to the global greenhouse effect, we are powering our electric grid with reliance on an *"energy transition"* technology. Burning natural gas is far less polluting than burning coal – if, and only if, we satisfy that all-important provision: methane cannot be allowed to leak.

The favorable medium-term economics of combined-cycle gas turbines (CCGT) fired by natural gas has allowed developing and developed countries alike to build them in droves. Referred to by their acronym, these power plants are responsive, efficient and quick to restart, stabilizing the electrical grid. They're so good that their low emissions are almost incidental. Before CCGT, baseload generation involved slow restarts, so any shut down was disruptive. Now, operating flexibility during low price periods avoids sacrificing low-cost wind or solar to keep expensive dirtier resources operating for when they're needed later. That pesky *curtailment* still happens if there's too much power!

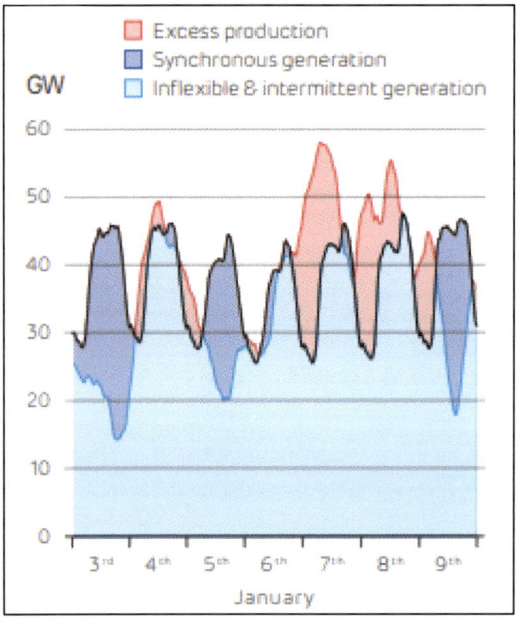


**Figure 8 – Renewables can provide almost all UK power averaged over this week, but their share ranges from 30% of demand to over 200% in specific hours. Synchronous generators, mostly gas-fired, flex up and down by over 30 GW in the space of just 8 hours (electricinsights.co.uk)**

This challenge will only grow going forward. To quote an expert:

> *"Taken across the whole of 2030, 34 TWh of electricity needs to be generated from non-renewable sources to satisfy [UK] residual electricity demand. At other times, there is 37 TWh of excess electricity production... What could be done to balance out this surplus and shortfall?"* [3]

The power grid varies its pricing over the course of a day, so large consumers such as factories employ teams dedicated to energy trading, deciding when to export or import power, getting paid enough to stop factories entirely at peak times. This approach has only gotten us so far. The big outstanding question of the day remains *energy storage*. There are many ways to store power – filling and draining reservoirs, compressing and expanding air, even lifting and dropping weighted blocks. Each has its own costs, responsiveness, and energy retention capability. None have been enough for us so far.

Although batteries have been around for a long time, and are now seeing astronomical improvements in materials and power electronics, they remain ludicrously expensive. Their improved energy-density and power-density makes today's batteries finally viable for day-to-day road and rail vehicles. But for grid-scale energy storage, it still seems nothing on the right order of magnitude is anticipated for the foreseeable future, given their prohibitive cost and scarcity of materials. Batteries and supercapacitors have only helped with short-term fast frequency response. Over an entire season, they don't store enough energy long enough. This means that transitioning to natural gas is often the best, or *only*, viable option to improve on an unacceptable status quo.

> Is it cheaper to provide standby power by arranging gas engines to sit ready-to-go, or to use ultracapacitors and specialist power electronics with battery banks? In the 2010's, both approaches started to be deployed together.

Better than combustion faraway is a capable gas distribution grid to deliver sufficient quantities of natural gas at suitable pressures directly to consumers. This allows them to set up *distributed generation*. Simply put: if fossil fuel is going to be combusted, it's best to do so at or near its actual point of use. This is because pipelines are more *efficient* than power grids at transmitting energy. In the absence of an all-renewable energy mix, and a power grid capable of handling that, responsive natural gas has rapidly become our power generator.

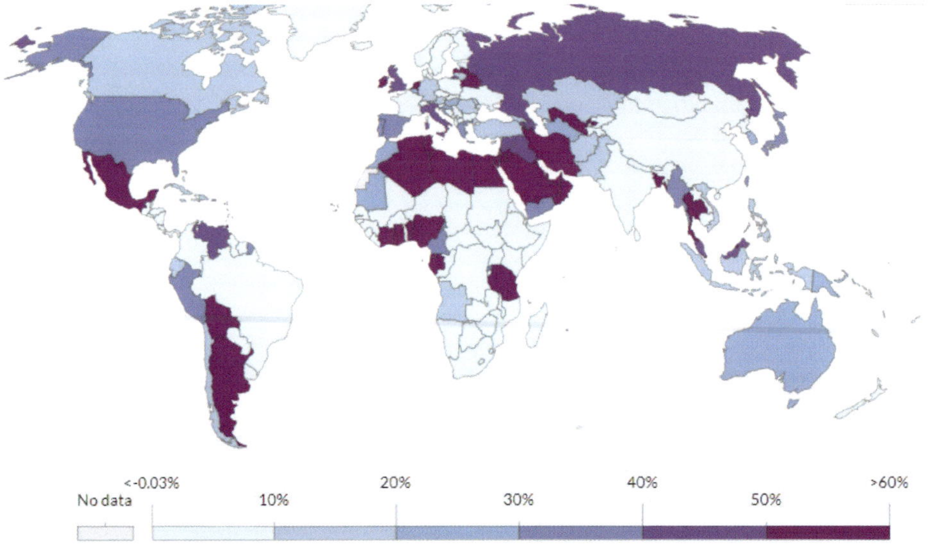


**Figure 9 – Electricity Generated from Natural Gas in 2019 (Our World in Data)**

The UK's energy policy is now officially to transition beyond natural gas to even more renewables, but it's vital we realize that as of 2020, almost every country powers its electrical grid on fossil fuels. The carbon intensity of this electric power grid is initially reduced by shifting away from oil and coal fired generation to natural gas power plants. Even if Australian gas is liquefied, shipped as LNG and gasified once again in China, that lifecycle analysis still results in better carbon emissions and air pollution than burning local coal. [4] It's tempting to consider pipelining as a political issue and shy away from defending its merits, but often economic imperatives reflect technical realities shouting: *"this makes a lot more sense than what we were doing before"*.

Some argue the only acceptable reality is to rely entirely on electric devices, implying some unspecified way to guarantee a grid can be balanced at all times – including substantial seasonal storage. Even if one adopts this position for argument's sake, we shouldn't let the perfect be the enemy of the good. At worst, natural gas is a useful transition technology, an accessible alternative to quickly replace more wasteful heavy polluters. Even the ambitious UK plan, weaning the country off fossil fuels almost completely as its declared goal, relies on natural gas to heat buildings for the foreseeable future, and on liquid fuels for many forms of transport. Decarbonizing isn't all or nothing.

Much of the world is still burning coal and oil where it ought not to be. Efforts exist at carbon capture and storage to reduce their carbon emissions, but that technology remains at the mercy of policy, requiring billions in consistent direct subsidies or emissions trading schemes to stay afloat, with no revenue stream to be gained overall. As the world shifts from coal and oil to natural gas, many users near(*ish*) plentiful gas fields cannot get enough of it flowing in.

As they attempt to make their transition, they find themselves struggling for natural gas supply, or limited by pipeline infrastructure reaching capacity. So the world is building natural gas pipelines. These crisscross the Black, Baltic and Mediterranean seas, linking North Africa, Russia and Asia to Europe. Many are politicized, technically unprecedented megaprojects. Russia's first gas pipeline to China sits on stilts to traverse Siberian permafrost, as was done for the Alyeska pipeline. This avoids the warm fluid melting permafrost around the pipeline and causing the soils to shift, potentially causing the pipe to break.

The digital revolution we are living through is knocking at the door of every economic sector, asking if doing things differently might make sense. Domestic consumers of electricity have become active participants, signing contracts to use batteries in their homes for smoothing out the wider grid, or exporting power generated from their rooftop solar panels. One potential avenue the *"internet of things"* could lead to is for devices that are plugged in at home to participate in the energy market by choosing good times to draw power, helping to balance the grid. Our washing machines could help save the world. It seems wise to take a holistic view, optimizing the power and gas grids in tandem, balancing both with whatever energy storage we can get our hands on. The UK's gas grid operators are looking at this link as seriously as electrical grid operators. A few who aren't have been politely told off by the government!

# A3 – ENERGY FOR TRANSPORT

**Figure 10 – Storage tanks holding aviation fuel at an airport (left) and an electric truck (right)**

For those destinations that aren't ever going to be linked by road and rail, how do we decarbonize the jet fuel tankers at the world's major airports? We rely on energy-dense liquid fuels. There are biofuel alternatives, but as we weren't careful about those, people took rash decisions that turned out to be terribly irresponsible, replacing some food crops with unsustainable palm oil for biodiesel or incentivizing the destruction of irreplaceable rainforests. In the US, corn oil is mandated to produce bio-gasoline, but that has its own issues. Research into algae as sources for biofuels is ongoing, but is far from mature. We remain a fair way from biofuels for aircraft: even if we technically manage, it is *scaling up* to adequate quantities that makes a technology revolutionary.

Smaller vehicles, including distribution within urban areas, are electrifying fast, so fast that in 2020 Norway was the first country ever to sell more electric vehicles than internal combustion engines. By 2030, many countries now plan an outright ban on the sale of all fossil fuel-driven vehicles, at least in the automobile sector if not initially in the haulage sector. This transition involves a greatly expanded electrical grid. Done right, there's an opportunity to exploit batteries of plugged-in vehicles as an energy storage asset for grid balancing.

Freight carriers along set routes were thought, by some, to be more amenable to hydrogen fuel cell drivetrains. In California and Japan, highway corridors were selected to form the backbone of hydrogen filling station infrastructure. But electrification by overhead gantry is also favorable on main routes – and the use of hydrogen for road transport is contentious, as we shall discuss later.

# A4 – ENERGY FOR BUILDINGS AND INDUSTRIES

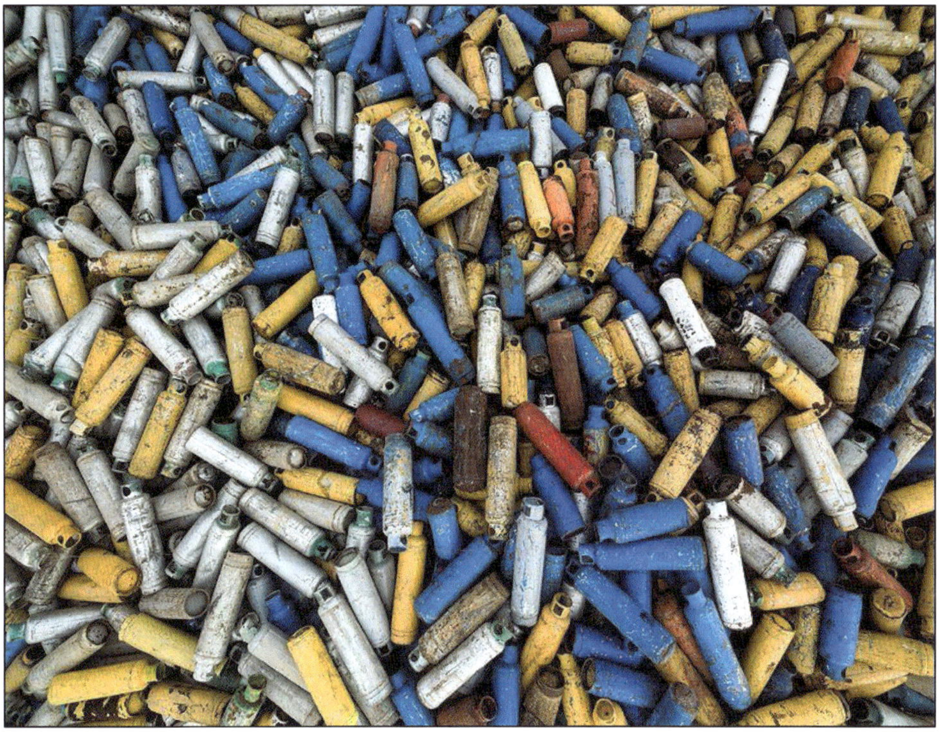

Figure 11 – Propane cylinders are still used in many countries, incurring expense and waste
Tomas Castelazo / Wikimedia Commons / CC BY-SA 4.0

I recall in my hometown of Alexandria, Egypt the effort expended to regularly carry cylinders of *"botagas"* up nine flights of stairs to tie into our cooking stove. That small-diameter pipe reaching our apartment brought great relief! And for those wondering why people didn't rely on electric hobs, it didn't cross our minds given the frequency of power cuts – though that has improved recently. Even with a reliable electric grid able to accommodate the extra load, many kinds of heating, cooling and industrial applications are not amenable to electrification. We simply can't electrify everything:

- *Inter-seasonal storage* – significant energy is stored across entire seasons to keep buildings warm in winter, and / or cool in summer, depending on climate.
- *Industrial users* – natural gas is used both as a feedstock to make products, and in furnaces to generate intense process heat at incredible temperatures.

We can reduce the heat a building wastes by improving its insulation. The UK's strategy is also to transition away, case-by-case, from natural gas boilers to several less carbon-intensive technologies. All new homes are soon to be equipped with electric heating, now greener than gas. Retrofits focus partly on decarbonizing using hydrogen, and partly on ground-based heat pumps. Heat pumps only make sense at scale; a central source feeds a district heating system of insulated pipes serving to circulate hot water around many buildings. Electrifying heating and cooling in buildings increases power demand and changes demand patterns. This needs additional generation, transmission and storage capacity.

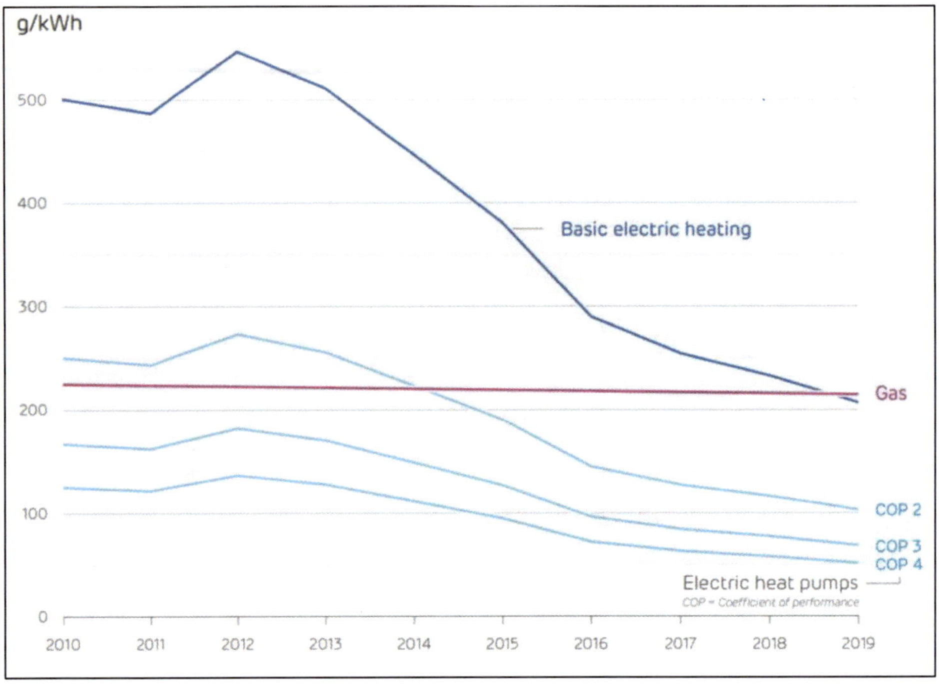


**Figure 12 – In the 2010's, the UK electrical grid's carbon-intensity became greener, making electric boilers cleaner than efficient gas boilers for the first time ever (electricinsights.co.uk)**

Thermal energy storage is typically done by heating up a molten solid, and is bound to help a little. Smart time-of-use tariffs can also shift heating demand away from peak periods. But the sheer magnitude of converting our building heat load from gas to electricity is proving to be a phenomenal challenge. Encouraging or mandating better use of ground-source heat pumps can help, as are simple measures such as installing double-glazing and insulation, but the lack of joined-up responsibilities for disparate buildings is a barrier. More to the point, the cost of ground source heat pump varies wildly by location. Not every city is built along a fault line, and the type of rock matters.

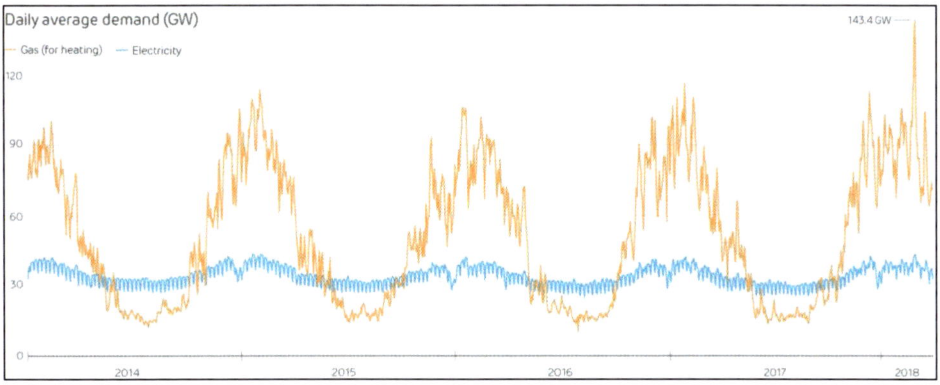


**Figure 13 – Daily average demand of gas for heating (orange) and electricity (blue) in the UK from 2014 until 2018, drawn against the same scale: gas pipelines and storage accommodate demand fluctuations orders of magnitude greater than the power grid (electricinsights.co.uk)**

There has been a push for feeding biogas into the grid, blending it into existing natural gas in careful proportion. Common practice was to generate heat and power on site instead of flaring at biogas facilities, but cleaning the biogas and blending it is a preferable use. It is one less conversion, and thermodynamics dictates that conversion entails inefficiency. Pilot schemes to export excess biogas rather than excess electricity succeeded, so this small energy source can do its part. But there is only so much organic waste and wastewater available to feed anaerobic digesters (AD's), so biogas can only ever be a small contributor. Even in the green and pleasant land of Britain, biogas can only ever heat 1% of homes.

Blending hydrogen into the natural gas grid might be more promising than biogas as a way of deeply decarbonizing our existing infrastructure. Just by mentioning it, we dive into a heated debate, with no shortage of disparaging remarks and name-calling. To many advocates of a green energy transition, the very idea of any kind of hydrogen seems to have become anathema. We discuss the production of hydrogen, how it is used, and how it affects pipelines in its own appendix.

# APPENDIX B – A HISTORY OF HYDRAULICS

## B1 – DOES HISTORY MATTER?

Having surveyed the present state of affairs, it is worth pausing to appreciate where we stand in the grand sweep of history. The origins of energy, hydraulics and fluids are intertwined with the story of the wider human condition. For better or for worse, countless energy transitions have already happened in the past. Perhaps this one will herald a brighter future.

It could be argued that distilling a concept to its pure essence leaves a fresh mind undisturbed by any pre-conceived notions, free to apply critical thinking and bring new perspectives to a subject. What I found myself is that history is edifying: the more I learn about the history of science and technology, the more I realize how incidental, incremental and almost haphazardly it all came to be.

It took people centuries to reach understandings and abilities that are now taken for granted. Even core concepts are glossed over by busy authors as clear, trivial or obvious. Yet if fluids and their flow were self-evident, people would have figured them out sooner! Superficially drilling formulas into our minds by rote memorization is never a substitute for real, deep understanding.

That understanding comes through experiments. But experimentation is often inaccessible, and where possible it feels like a frustrating exercise in futile dead ends. Looking through the embarrassing history of *anything* serves to provide some reassurance. We are likely to struggle as we strive to understand, as we strive to make something better, and as we strive to overcome insurmountable obstacles. Blockers will place themselves in our path. That struggle, enduring all kinds of challenging circumstances, is the hallmark of true innovation.

This review of the history of pipelines is presented as a consolation. Although a reader who feels burdened reading through the history we present here should feel free to skip it entirely, it provides context and understanding for the human stories that led to the professions of pipelining and simulation.

# B2 – ANCIENT PUBLIC WORKS AND INVENTIONS

## Ancient water channels

For the field of hydraulics to have taken so long to formalize into a conceptual framework, complete with underlying principles, is something of a surprise. Water conduits appear very early on, almost since the dawn of written history. The Romans were famous for their long-distance *aqueducts*, and they were by no means alone. Many civilizations – European, African and Asian alike – built tall bridges, dug long subsurface trenches and excavated tunnels bored deep underground, through which water flowed perennially to farms and cities. Even on their isolated continent, the pre-Columbian Mayans exploited water pressure for flow: it's intuitive that water flows downhill, but only experienced engineers know to design a water passage to get narrower as it progresses.

In the ancient periods of history, the written record was sparse and uncertain. Undoubtedly, we have reports of open channels transporting water in India, but little of that remains, or at least little of it is documented well enough to talk about in any detail. Almost all civilizations, of course, dug canals for irrigation and water supply, but those are not our focus of interest in this book.

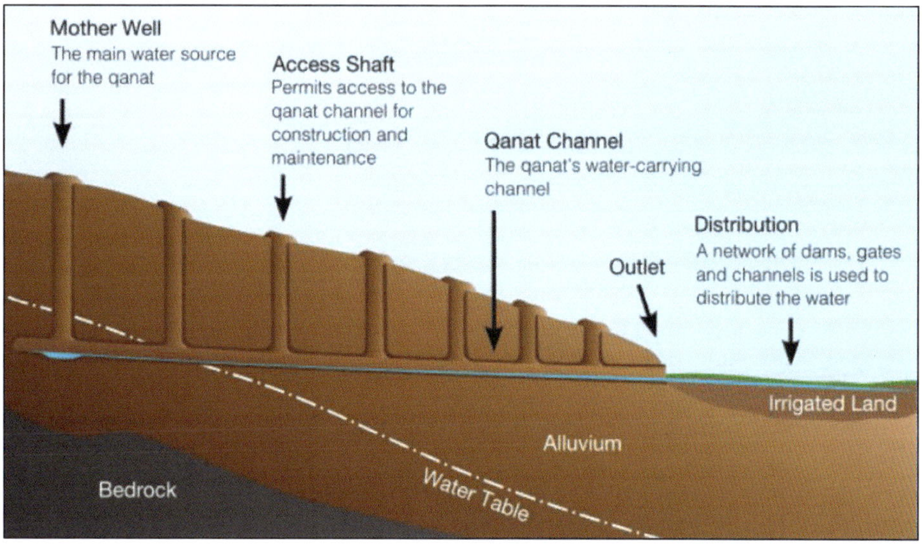


**Figure 14 – The ancient principle of an underground *Qanát* system**
**Samuel Bailey / Wikimedia Commons / CC-BY-3.0**

What is a little more interesting is an ancient Persian system of *underground* channels. Their principle came to be known as a *Qanát*, allowing arid regions to flourish. A tunnel is dug at grade from a point at which the water table is higher than the elevation at their delivery point. At regular intervals, access shafts are carefully cut vertically to allow for maintenance and, importantly, ventilation. Water within a confined space releases its dissolved gases, some of which are flammable. The water is distributed for irrigation, or carried to huge cisterns, reinforced at their bottom with metals to carry the structural load exerted by tons of water. From there, smaller open channels distribute locally.

Lots of ancient Persian technologies are alive and well in that part of the world. The *windcatcher* was an architectural feature achieving natural ventilation by carefully exploiting air currents and temperature differences between rooms. Combined cleverly with an underground channel (a *Qanát*), windcatchers kept *ice-houses* cool, the building's thermal mass performing what we call diurnal energy storage. But other than this clever arrangement, drawing air along by the flow of adjacent water, the art of ventilation and of controlling temperature and humidity remained the domain of buildings and architecture, far removed from hydraulics. Craftsmen could do great things, but they did not quantify these phenomena, nor did they worry about *why* they happen as scientists do.

Channels flowing by gravity at atmospheric pressure do not immediately seem relevant to understanding pipeline hydraulics. They are more a testament to social organization, surveying methods and materials. It is quite possible to dig a tunnel for water to emerge out of the side of a hill without worrying at all about how much of it will flow out. Its flow would change anyway if the aquifer is fed by meltwater and rainwater, so why worry about what they can't control?

What is interesting about the *Qanát* is that, like the Roman aqueduct, over the course of middle ages it proliferated far beyond the Persian sphere of influence. In fact, though it's used in the modern language today, the term itself is a loanword. Its traditional principle was developed to find new applications as technology emerged to raise water.

## Ancient water pipelines

Surprisingly, the ancients didn't just hew channels out of stone. Pipes made of clay, lead and copper (even stone!) were widely used in the Middle East, China and across the farthest reaches of the Roman Empire. We see those pipes at archaeological sites, and a few Greek and Roman books on the topic survive.

The Ancient Egyptians installed copper pipes for a funerary complex within a pyramid in Abu-Sír. This was plumbing for washing rituals, not a real pipeline.

In the realm of the living, our first glimpse at long-distance pipes might be at Samos. We may never have found that aqueduct, were it not for Herodotus describing its mile-long pipeline. It dates to the 6th century BC. An engineer known as *Eupalinos* extended a piped water supply across rugged terrain to the capital of the island. This included tunnelling straight through a mountain, and digging ditches to link the mouths of that tunnel to the source at one end and the town at the other. Within this ditch (including the portion tunneled through the mountain) they buried pipes. The purpose of the ditch was to provide a bed for the pipes that kept the pipeline at a consistent but ever so slight downhill gradient along the length of its entire route. *Terra cotta* is Latin for baked clay, and could not fare very well in tolerating much pressure.

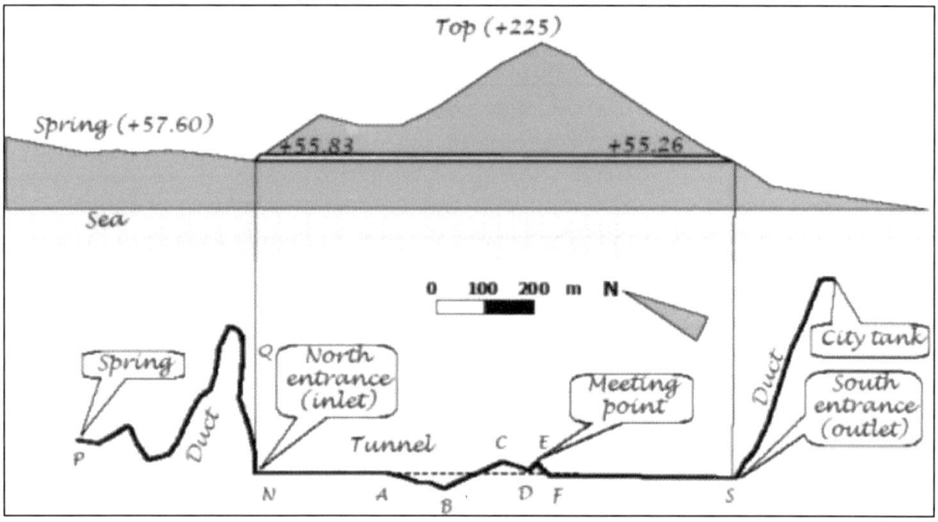


**Figure 15 – The clay pipes of the Samos aqueduct withstand little pressure, so they buried them in a trench dug at a careful grade through a mountain. (eupalinos-tunnel.gr)**

In the subsequent centuries, the Greeks mastered water pipes down to a tee. In 2000 BC, for the palace of Knossos in Crete, terra cotta pipes tapered to increase water pressure, preventing sediment build-up. Handles on the outside were tied together with ropes, and all joints were sealed to prevent leakage. [5]

At Pergamon, archaeologists found a long pipeline with an inverted siphon. Water was carried from a source 1 200 feet above sea level, taken down across two valleys, then piped back up again to reach a citadel, the summit of which is only 130 feet below the height of the spring. Archaeologists traced the route of this pipeline by identifying the stone sleeves in which the pipes were set. The system must have been extremely well-maintained to hold so much pressure. The aqueduct consisted of pipes made of terra cotta, with inner diameter 0.13 m and length 0.6 m. The base of the siphon was 25 m below the inlet, a head of water that imposed a static pressure several atmospheres above ambient pressure (2.5 barg). [6]

Figure 16 – Parallel pipes made of terra cotta at Pergamon (RomanAqueducts.info)

The Ancient Greeks got some laws of nature right. *Archimedes* experimented with water in Sicily, founding hydrostatics in 150 BC. As an engineer, or in the language of the time, an *"architect"*, he was a jack-of-all-trades. His remit began with machines of war to help his Sicilian Greek king fight off the Romans, and extended to all sorts of wondrous buildings and useful devices.

He invented a famous screw pump, which remained in use by farmers in Egypt as recently as the mid-20$^{th}$ century. He described how a floating solid object displaces an amount of water commensurate to that solid's density, the same principle of buoyancy letting heavy ships float, so long as their displaced volume is large enough to cause an upthrust carrying their weight. He applied this to perform a non-destructive test on the purity of that Greek king's crown, foiling the cunning scheme of a cheating goldsmith who hid silver in its gold.

But some Greek philosophers also left us with wrong ideas that took centuries of experiment, analysis and debate to overturn. *Aristotle* insisted that air has no weight, which took longer to overturn than the acceptance of heliocentrism!

People have long known that sealing a fluid in a conduit makes it capable of withstanding pressures above atmospheric. Rather than open channels, they often chose to construct transmission and distribution systems as covered channels, to protect them from contamination and avoid evaporation. But masonry doesn't hold much pressure. So the level of the water matches what we now call the *hydraulic grade line*. They built gentle slopes all the way down. But a gentle slope involves major infrastructure and engineering megaprojects.

Figure 17 – The Romans built the *Pont du Gard* in the 1st Century CE

What if a valley is too steep to cross by building a massive viaduct? The Greeks taught them that water flows uphill against gravity if it just came down a bigger hill. To do that, they used *pipes*, whose thick walls made of strong materials are formed to make a circular cross-section, maintaining significant pressures. In Iberia, a well-preserved example of a Roman aqueduct is at Cádiz. For a portion of its route extending over 3.5 km, it descends 100 m into a valley and emerges from the other side. For this crossing, rather than building high bridges for a channel, they constructed a pipeline to create an *inverted siphon*, holding a 46.5 m head of water, imposing 4.5 barg of static pressure, which probably explains their choice of costly masonry (rock!) as a material.

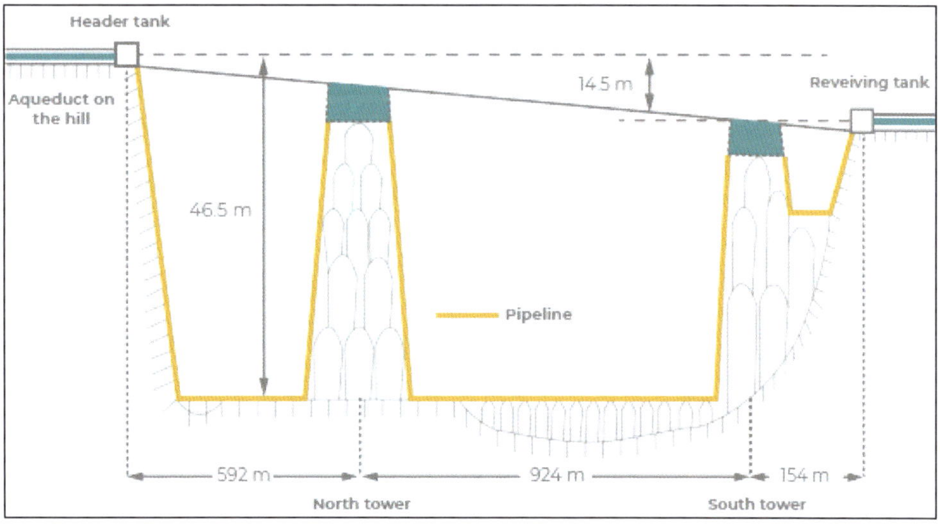


Figure 18 – The inverted siphon at Cadiz, Spain was one of many built by the Romans. The pipes were not strong enough to withstand traversing the entire valley, so crossed part of the valley on bridges as a pipeline aqueduct (Adapted with kind permission from RomanAqueducts.info)

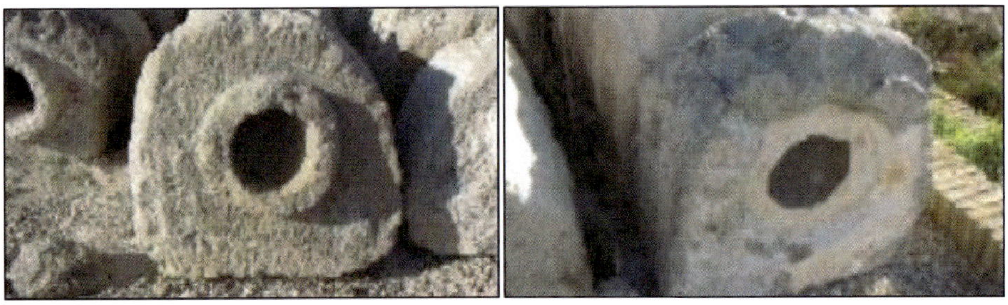

Figure 19 – Masonry pipes of the aqueduct at Cadiz, Spain (Credit: RomanAqueducts.info)

As was common practice in Roman Spain, the pipelines at Cadiz were made of masonry. Lead was malleable and easier to work with, but available as an affordable by-product only near places with silver mining. Roman craftsmen chose to build most pipes of lead – PLUMBUM. In English, we still say *plumbing*. Lead poisoning was an issue, and they knew the risks, though it was probably the kitchenware more than the pipes. They had a hard water supply, a peculiar quirk of Rome and preferred in taste by the Romans wherever they tested water for its quality. Hard water deposited its minerals onto the inner surface of their lead pipes, coating it with a protective layer. The public health risks of lead pipes, of course, still live with us today, recently affecting Flint, Michigan. Nonetheless, terra cotta pipes were certainly the safer option. We know that the Romans were well aware of this public health risk, because they told us! [7]

> *"Clay pipes for conducting water have the following advantages... if anything happens to them, anybody can repair the damage. Secondly, water from clay pipes is much more wholesome than that which is conducted through lead pipes, because lead is found to be harmful..."*

The oldest book on engineering still in existence today is DI ARCHITECTURA. Its original illustrations have been lost, making it notoriously hard to decipher in places, an apt reminder that drawings help us a lot when discussing technical matters. Its author, *Vitruvius* (fl. 15 BC) worked with the emperor Julius Caesar himself, dedicating the book to the emperor's adopted son, Augustus Octavian. Despite their famous constructions, the Romans almost never wrote on engineering, illustrating the disdain of their ruling classes towards technology. Vitruvius wrote his book striving to change that, but to no avail. In it we find all kinds of interesting nuggets. He seems to be the earliest source to mention Archimedes jumping out of a bathtub (*Eureka!*), a probable fabrication but one that has proven effective in popularizing science throughout the ages.

He lists standard lead pipes in use at the time. Pipe size was expressed not as a diameter but a *circumference*. Wall thickness was implied by the weight of a lead sheet from which pipes of that width are rolled and welded at the seam:

> *"The pipes should be cast in lengths of at least ten feet. If they are hundreds, they should weigh 1 200 pounds each length; if eighties, 960 pounds, ..."*

That seam presented a problem: it split apart, imposing a limit on pressure far lower than that of the actual material of the pipe wall. Methods to manufacture seamless pipes were not invented for many centuries, restricting the pressure. To protect against a pressure surge, they installed relief and ramped up slowly:

> *"...in the venter, water cushions must be built to relieve the pressure of the air... a strong current of air is generated in an aqueduct which bursts its way even through stone unless water is let in slowly and sparingly from the source at first, and checked at elbows or turns..."*

He tells us that they positioned separate connected tanks at different heads, so that private villas don't deprive the wider public from a basic water supply.

**Figure 20 – An urban distribution tank (CASTELLUM) at Nîmes, France**
**Pipes radiated from this central reservoir, fed by an aqueduct of square cross-section**
**Przsak / Wikimedia Commons / CC BY-SA 3.0**

Once the water has traversed the aqueduct and finally arrives at the boundary of the city it is intended to supply, it was carefully stored and distributed:

> *"From [a] central tank, pipes will be laid to all the basins and fountains. From [a] second tank, to baths so that they may yield an annual income to the state. And from [a] third, to private houses, so that water for public use will not run short, [and] in order that individuals who take water into their houses may by their taxes help to maintain the conducting of water by the contractors."*

Those contractors were the *water-men*, some of the first pipeliners in history. And they earned themselves quite the wretched reputation. A century later, in 96 CE, *Frontinus* was commissioned by the emperor to write DE AQUAEDUCTU, a two-book official report on the state of Rome's aqueducts. In these he wrote extensively and in great detail about who was ripping off the Roman state and its populace of, by now, a million people, and exactly how they were doing it. The cabal of water-men had formed elaborate criminal enterprises to short-change the citizenry, oversizing pipes which flowed in to supply themselves, undersizing pipes which flowed out to meet the demands of their customers, and manipulating the hydraulics of Rome's water pipelines with impunity. They would, for instance, tap into a reservoir at a lower point than they had been allocated, to get more pressure and therefore more flow than they pay for. One famous trick was quite literally manufacturing loopholes to circumvent agreements, tapping into mains water supplies with an orifice of the proper diameter, then attaching that orifice to pipes that are much larger in diameter.

It had all gotten completely out of hand. The city of Rome was consuming *six times* as much water per capita as we do in modern times! Even accounting for the lack of outlet valves (taps / faucets) in people's homes, the game was up. When the Romans did what they did best and tightened up their bureaucracy with better procedures to enforce laws and regulations, they resorted to *stamping* the water pipes! A stamp on every pipeline proved its ownership and traced its manufacturer, allowing inspectors to check against their records and sign off that the size and route of a pipeline is in good order.

Figure 21 – Roman pipes inscribed with manufacturers and owners
Wolfgang Sauber / Wikimedia Commons / CC BY-SA 1.0

Despite those efforts, the Romans never quite managed to get their water pipelines back under control. The men charged with inspecting that infrastructure – hundreds of slaves reporting to men like Frontinus – were themselves corrupt, or at best powerless. The men stealing the water were mostly wealthy landowners tapping into aqueducts to irrigate their farmland.

It is remarkable that despite the Romans' lack of understanding of several key concepts of liquids flowing through pipes, that did not stop them building cross country pipelines from sources to cities, plus distribution pipework within the cities. It proves that the barrier to entry for a technology to become useful is often societal, not scientific. Their pipelines were not perfect, but they worked. Despite being incredibly wasteful, Roman pipes served their purpose to deliver water for drinking and for irrigation, quite literally sustaining their civilization.

# Ancient steam baths

A bathhouse in any part of the Roman empire hosted rooms of various conditions of temperature and humidity: dry and hot, wet and hot, dry and cool. The tradition continued in some places throughout the medieval period, and after all these centuries it can still be experienced today in several cities. The bath-goer is taken from a sauna, to a cool pool, and so forth, in a careful order.

Vitruvius describes how to build and operate this HYPOCAUST, including how fuel could be conserved by building the hot room (CALDARIUM) adjacent to the warm room (TEPIDARIUM). This is intuitive to us today, as we know heat transfer scales with temperature difference, but that sentence could not be phrased in that way for most of human history. A central furnace raised steam. Temperature could be adjusted by one of two approaches. Some bathhouses passed hot exhaust gases from the furnace through an underfloor cavity beneath the space. Others used the exhaust to heat an intermediate fluid with high *heat capacity* – such as oil – through pipes embedded within the building. Valves admitted a varying amount of steam into each room, to blend with the air and raise its humidity. A bronze ventilator adjusted how much steam was released from the domed ceiling.

Like their hydraulic projects, the Romans did not concern themselves with the science underlying their technology, relying on rules of thumb and experience to control how hot or cold a room is, and how humid its air is. The ancient Greek philosophers had put forward certain theories about what matter is composed of, and tried to explain its properties and interactions, but the Romans were not particularly interested in those theories. It was in the Greek tradition that philosophy – including natural philosophy – was pursued, even after the fall of the western Roman Empire, and the long, slow decline of Byzantium. As the saying goes: anyone can build an aqueduct, it takes an engineer to build one that barely stands. I'll add: it takes a scientist to say why!

**Figure 22 – A Roman lead pipe in Bath, Britain. That thick silver ridge is the soldered seam which needless to say, is very different to how modern day pipes are made**
**Andrew Dunn / Wikimedia Commons / CC BY-SA 2.0**

# Ancient salt, oil and gas

There is a surprising connection linking ancient salt mines with pipelining and today's oil & gas industry. Salty brine naturally emerges to the surface in some parts of the world. In a select few locations, combustible fluids also emerge to the surface. It should not come as a surprise, then, that people used all of these since ancient times.

The ancient Greeks knew a recipe for a napalm-like substance ("*Greek fire*"). By the Caspian Sea, they saw people dipping arrowheads and spearheads in oil, setting them alight, and launching them at their enemies. The Zoroastrian inhabitants of that region used this oil and gas to light fires for ritual worship. This tradition continued around Baku in Azerbaijan, well into the middle ages.

In the inland Chinese province of Sichuan, salt was hard to come by so far from the sea. It was a sought after commodity preserving food in an age before refrigeration; the salt trade was valuable, taxed all over the world. Sichuan's need for salt was not unique. Nor was the response of its market in striving to supply that sought after commodity. Natural pools of brine supplied some salt: brine was boiled off in salt pans heated by furnaces. But people needed more.

**Figure 23 – Brine being boiled down in the salt pans of Zigong, China, the vast scale of the region's industry made sources of wood scarce for fueling its furnaces**
**Phreakster 1998 / Wikimedia Commons / CC BY-SA 2.5**

In the 3rd century BC, a hydraulic engineer and administrator named Li Bing, famed for his huge public works projects, discovered those pools originated underground. Around the town of Zigong between the capital of Chengdu and the Yangtze river, he ordered the world's first brine wells to be dug. At first the wells had wide mouths, akin to an open pit. Some went 90 meters deep. [8] These relatively shallow wells of ancient times are known as *shaft wells.* [9]

Sometimes, the workers digging the wells would fall ill. Occasionally, a tremendous explosion would kill an entire crew, or spit flames out of the bore holes. By 68 BC, two wells in Sichuan and neighboring Shaanxi, became infamous as sites where this *"evil spirit"* rose out from the underworld. The governors of the provinces would visit every year and make annual offerings. By 100 CE, the workers understood the disturbances were caused by an invisible substance. They identified the holes where it emerged, set it alight, and used it to cook. As the population grew, the demand grew. They drilled thousands of wells. As they developed their drilling techniques, their wells got narrower and deeper, drilled using iron-tipped bamboo rigs and shoring up the deep shafts. They extracted brine by inserting a length of bamboo tube with a leather valve at the bottom, lifting it up by an attached cable, and pouring the brine into a tank. From there, brine flowed through bamboo pipelines to the boiling houses. [8]

As the industry grew, sources of wood to fuel its furnaces became scarce. *"Soon they learned to insulate bamboo tubes with mud and brine and pipe the invisible force to boiling houses... by 200 CE they had iron pots heated by gas flames – the first known use of natural gas in history."* In 347 CE, wells were identified where the flammable substance known as *"burning water"* emerged with the brine. Natural gas in this region is *sweet*, devoid of poisonous contaminants, so it neither stunk nor poisoned workers. [8]

Figure 24 – Brine in bamboo pipelines (Credit: Needham Research Institute nri.org.uk)

Figure 25 – Gravity-driven liquid pipelines made of bamboo, raised presumably because a significant siphon would cause the bamboo to burst due to over-pressure

# B3 – MEDIEVAL TECHNOLOGY AND SCHOLARSHIP

## Progress and regress

How medieval philosophy and technology nudged (wholly wrong) ancient ideas towards (marginally less wrong) early modern science is not an uncontroversial topic. It is clear there was substantial progress in certain respects, while in others established opinions stagnated, or people ceased working on a topic altogether. If we ask: did thinkers of subsequent centuries build on what was there, or start over from a clean slate, the answer is both. For most of history, the physics of fluid flow utterly baffled those who made fluids flow for a living.

At first, the Roman idea of *liberal arts* persisted, in contrast to the *servile arts*. Free gentlemen did not stoop to the level of mere artisans and smiths, reducing themselves to manual pursuits, sullying their refined minds with lowly trades. That changed dramatically. People started measuring flows in conduits, trying to quantify it by rudimentary means. A few started to ask why water flows in the quantities that it does through a channel of a certain size. Diverting flows arbitrarily came to be seen as wasteful, and unfair. And natural philosophers struggled to reconcile their theories with new technology – the hydraulics of ingenious contraptions such as water-clocks, and machinery such as pumps.

An atmospheric channel is a feat in civil engineering – moving earthworks to carve conduits, hewing tunnels out of solid rock, constructing bridges and their supporting structures. It is eye-wateringly expensive to build and maintain. Even if it is covered to reduce evaporation, it must maintain a careful descent, because the water flows along a channel by the force of gravity alone. Since there is no pressure difference anywhere, it is nothing at all like a sealed pipe – in a hydraulic sense anyway. If a pipe is everywhere at atmospheric pressure, it is probably only part-full, and people have always referred to it by a different, distinct term: a culvert. Because we are concerned here with pipelines, neither channels nor culverts are of immediate relevance in terms of their fluid physics.

Looking for an inverted siphon to get a watercourse across a deep valley, that situation which warranted proper pipelines peppered across the ancient world of Greeks and Romans, we are hard-pressed. The inverted siphon seems to have disappeared where other technology flourished and advanced. Why?

We must leave for another book the fascinating but quite complicated topic of what exactly happened to ancient pipelines. Why some of those pipelines were maintained and improved throughout the medieval period, while others were abandoned altogether. In another book, the author can elaborate in more detail how the development of ancient ideas on natural philosophy set the scene for what we call the renaissance, and eventually for modern pipelining. However, we shall bring forward a couple of short examples that serve to illustrate core ideas about the hydraulics of pipelining before moving swiftly on to later times.

In the first example, water is pumped through pipes by exploiting pressure, or to be more precise, a negative gauge pressure, i.e. a partial vacuum. This is a better way of raising a fluid than having a scoop dive in and dump it at a higher elevation – it is not a *"chain pump"* but something a lot more like a true pump.

**Figure 26 – Al-Jazari's water-powered pump manuscript (left, Topkapı Palace Museum) alongside a recent half-scale reconstruction (right, London Science Museum)**

Al-Jázari (d. 1206 CE) improved on earlier work by Ctesibius (d. 222 BC) with several original features in a twin-cylinder reciprocating pump. Powered by a water-wheel, its crankshaft converted rotary motion to reciprocating action. This is considered to be the first known suction pump: instead of having to be placed in the water, the machine can be positioned more conveniently above the water level, with suction pipework extending down to draw the water into its vacuum. Today we refer to this arrangement as a net *negative* suction head.

The second example was what to do about open channels being essentially uncontrollable. Although glimpses of ancient practices developed in medieval times, we see pipes primarily in urban settings. By and large, the cross-country supply of water from sources to urban centers came to rely on gravity fed channels, overground or underground. It's not possible to shut off a channel: it would simply overflow. One can only *divert* water, otherwise it goes to waste.

This raised the question of seasonal balancing: surplus water in winter to supply the dry months through the summer. Storing the immense quantities of water necessitated huge, expansive storage basins. And if the rain in a region would possibly fail for years during a period of drought, longer-term stores of water were needed. How much to accommodate in a reservoir is a pretty straightforward calculation. It warranted humongous constructions. Interconnected reservoirs, settling tanks and underground cisterns were built all over the medieval world.

The third and final example was distributing water fairly through channels. This became the topic of several books. Users were being charged prices for apportioned hours of the day or night. What price was fair became a complicated question. Experts in law and jurisprudence had a stab at weighing up competing commercial and public interests of all parties. But it is among the mathematicians that we see early tentative attempts to improve on the Roman method of relying on cross-sectional areas rather than just apportioning hours.

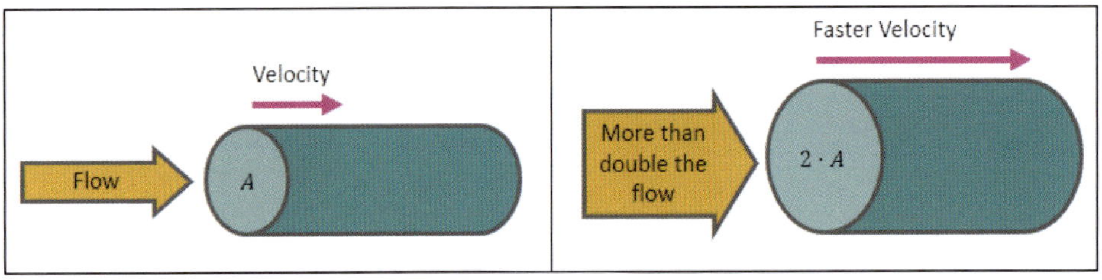


**Figure 27 – Flow is not proportional to cross-sectional area: larger conduits flow faster**

Several mathematicians started to *quantify* the physics of motion – including flowing water – with numbers, expressing what natural philosophers could not yet say. *Ibn al-Bánna* (d. 1321) of Marrakech spent his career popularizing mathematics with practical applications accessible for a broad readership. One of those applications was the distribution of water in conduits. He rightly remarked:

> *"Those who apportion water collect it in one location. Then, they make the ratio of orifices through which water exits as the ratio of the parts they wish to divide it into, believing that such is a just division. But that is not so, because through a larger orifice water flows faster, taking a greater proportion of water. Moreover, [of] the channels distributing water, a lower one draws more water,*

*thus carrying more than the [fair] proportion. Even if they are equal in distance and descent, the one with [less] bends draws more water, thus carrying more than the [fair] proportion. A single branch amongst them might have several of these [features], thus carrying much more than the [fair] proportion."*

Sadly, it seemed nobody picked up on this astute observation. It is one of the earliest formulations we could find clearly expressing the effect of a conduit's properties on the quantity of fluid that flows through it, but unfortunately this topic seems to be poorly researched, so I would not be surprised if a reader got in touch with other earlier and later sources striving to quantify flow rates.

# Medieval oil and gas

Well into the 1800's, a preferred source for the best lighting oils was seals and whales. These were priced beyond the reach of most, being an expensive luxury commodity. For its clean, long-lasting burn, whale oil was favored over vegetable oils, and those were favored over tallow from animals. But what of petroleum wax? The oil reserves around Baku, while not the only ones in the world, were most remarkable. Exploited since ancient times, they developed on a commercial scale in the middle ages, benefitting greatly from the expanded trade networks. The oil revenue of whoever ruled that region was substantial. Chroniclers and travelers like *al-Masúdi* (d. 956) tell us how Baku oil was transported far and wide by ships and caravans at great expense. It was used as a fine commodity, for perceived medicinal virtues, lighting lamps and various weapons of war. [13]

As for many technical concepts, Arabic borrowed a loanword, the Persian term *naft*. Crude oil was called *black naft* (whatever color it may be in its natural state), and distillates were called *white naft*. The distillation process produced successive fractions, and those fractions were described by some founding names in alchemy.* *Al-Rázi* (d. 925) elaborated in detail the process for distilling crude oil, and how to use its products: petroleum distillates. [13]

Baku had so much oil that much of it emerged to the surface of its own accord. By the 1200's, wells were being dug to reach the level of the oil underground. *Marco Polo* (d. 1324) reported a hundred shiploads of it were taken at once. [13] In 1572, Safavid

* Experimentalists insisted on verifying claims, and despised Hermetics for sullying their reputation with quackery. Words like *gibberish* (Geber-speak) fell out of this quagmire. Use of *alchemy*, *"chymistry"*, et cetera, was inconsistent.

Baku was visited by agents of a British trading company interested in its oil. It was characterized according to its color, the best being yellow oil. [10] Further west, near Mosul, seepages of oil were also long known at the surface, and *al-Antáki* (d. 1597) describes the process for its distillation. They found they could ignite light fractions by motion alone, without a flame. [13]

**Figure 28 – Bamboo pipes raised on trestles, transporting brine to boiling sheds (Image Credit: Paul S. Noll, paulnoll.com/China/Excursions/Zigong-salt-pipes.html)**

Meanwhile in Sichuan, by 1100 CE, a bureaucrat named Su Dongpo *"built sophisticated bamboo urban plumbing. Large bamboo water mains were installed in Hangzhou and Canton. Holes and ventilators dealt with blockages and air pockets"*. These bamboo pipelines were deployed on a wider scale, irrigating farms and supplying water to villages and homes. [11] However, the deterioration of bamboo requires frequent maintenance, replacing a pipe after only a few years. This deterioration is slowed by the salt content of brine, so it is in that application that bamboo pipelines found the most lasting success.

The salt industry of Sichuan was profitable only because of local brine being cheaper than heavily taxed salt from coastal areas, so although it did develop tremendously, it didn't spread beyond the immediate region. [9] For this reason, we do not see bamboo pipelines in any great quantity except in that particular region, serving the salt industry.

At shallow levels, brine is of questionable quality, laden with sediment and only yielding a little salt. By 1100 CE, to reach better brine, salt producers developed percussion drilling. In 1071, Wen Tong wrote that:

> *"[two decades ago], the local populace began constructing wells by drilling into the ground and installing bamboo pipes. These borehole wells are used to extract [brine] which is then boiled and refined into salt. [They] have become wholly skilled in this method."*

When a government ban on new wells was relaxed, thousands of boreholes were drilled. Around the same time, Su Shi narrated the technical aspects of those drilling techniques. Bamboo tubes served as the walls of the well, keeping groundwater from contaminating the brine at the bottom. They reached depths of 120 meters. [9] Lifting brine from deep wells is too heavy for manpower, so drums raising brine to the surface were rotated by animals. [12]

Eventually they drilled new wells nearby, over 240 meters deep, specifically to tap into the gas reserves they had been coming across incidentally since ancient times. Sources suggest different dates for when this region linked gas from dedicated wells to end users using bamboo gas pipelines; it was certainly no later than the 1500's CE. To convey gas to the salt works, they used hollow bamboo tubes as the world's first gas pipelines. [11]

In the 1000's China started using gunpowder, one of the first major industrial applications for a potassium *"rock salt"* known as saltpeter. In the 1100's, while the first wave of European Crusaders was sacking Constantinople instead of conquering Jerusalem, Arabic literature starts mentioning the secret Chinese powder. [11] Combined with distilled petroleum, this gunpowder fundamentally changed the nature of warfare as new weapons emerged. It also transformed hydraulics and gave rise to the pressures associated with modern pipelining.

# Gun barrels reining in the air tight

> *"The earliest approximations to pressure-tight containers were the barrels of guns, and historically these were never satisfactory and frequently failed. A list of people killed by the accidental bursting of guns, from King James II of Scotland downwards, would be long and impressive..."* [14]

Gunpowder was brought to the Mediterranean world by the interconnected minds and markets that braved the many silk roads. Those trade routes were afforded new importance after Genghis Khan sacked Baghdad. Travelers through these lands, now ruled by assimilating Mongol dynasties, must have brought the first formulas with them. *Al-Rammáh* (d. 1295) writes in his book of *"Chivalry and Military Contraptions"* how to blend gunpowder with petroleum. This process produced another napalm-like substance, an incendiary that they launched from wooden gun barrels.

For all intents and purposes, steel is just iron with a certain content of carbon. That carbon is imparted to it during the process of smelting to extract it from its ore. Cast iron has 2-4%. Wrought iron has 0.1%. Steel has something in between. Throughout written history, iron and steel were made by many processes, tightly held as industrial secrets. Steel blades were forged in China for well over a millennium, and swords from India were held in high esteem. Over the centuries, methods and prestige passed on to Damascus and beyond. While a number of authors wrote on steelmaking, it was never a cheap material, as processes could not produce it at high quality in large quantities.

The proliferation of waterwheels in the medieval period allowed more air to be blown in successively better processes, producing better iron. This eventually let people cast more impressive gun barrels. Over the next centuries, in obscure circumstances, advancements in all these domains led to *cannons* appearing, creating tremendous pressures by using gunpowder in explosive proportions. One clue is that this happened in the western region of Iberia, where petroleum was scarce but there were vast reserves of saltpeter.

However they came to be, huge bombards were used to knock down the walls of Constantinople in 1453, and cannons were used in a lot of warfare during that period, including the fall of Granada in 1492. In that same year, ships first sailed to the Americas and returned with so much silver that prices exploded too (hyper-inflation). These events are seen by most historians as marking the end of the middle ages.

Figure 29 – The Ottomans made huge bombards soon after the discovery of gunpowder. These pressure vessels drove metallurgists to improve their methods.

The need for reining in pressure had arrived in force, and metallurgists struggled to keep up. Containing the high pressures of expanding gunpowder became a preoccupation for ironsmiths, whose forges weren't merely foundries but versatile manufactories. Skilled craftsmen worked iron to make machines. Those included the first pressure vessels, for the first engines and pipelines. *

# Raising water from flooded mines

Flooded mines were a stubborn pain everywhere and throughout the ages. In his extensive chronicles, *Al-Qazwíni* (d. 1283) wrote of a place in the *Maghreb*:

> *"There are underground mines in which many people are always working. When they descend 20 cubits, water appears. The Sultan installs waterwheels, and water is raised in three stages. He installs a wheel down in the mine at the water's surface. Water is lifted and discharged into a large tank. Another wheel is installed on this tank. It lifts water and pours it into another tank. On this tank a third wheel is installed. It lifts water and discharges it on the surface of the ground to irrigate farms and gardens."* [13]

In China, *Wang Zhen* (d. 1333) published an illustrated agricultural treatise featuring horizontal and vertical watermills, the technologies preceding axial and centrifugal pumps to come. He also built waterwheels driving other machines, such as bellows, an innovation that facilitated forging better iron.

---

* The term for pipes in Arabic manuscripts (*unbúb*) was the same one they chose for gun barrels. It is still used for pressurised gas canisters and pipelines.

**Figure 30 – Vertical and horizontal waterwheels from *Wang Zhen*'s treatise**

The chaotic non-hierarchical way engineering was practiced had its upsides. Many acclaimed talents throughout history were persecuted, yet sought no permission and made no apologies. *Leonardo Da Vinci* (d. 1519) left us striking illustrations that ponder the turbulence of a streaming liquid, as if to invite those who followed to address unanswered questions lurking in these murky waters.

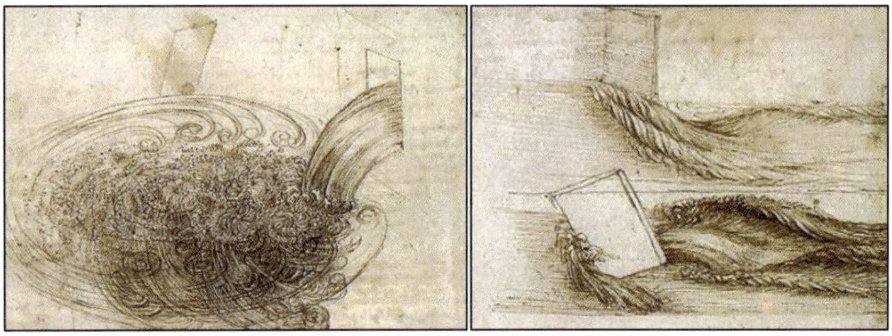

**Figure 31 – Studies of water passing obstacles and falling by Leonardo da Vinci**

Europe's first book on the topic of mining, published in 1556 by *Georg Bauer* (latinized Georgius Agricola), literally illustrates the age-old problem. That pressing need to extract water from flooded mines offered both a humanitarian and financial imperative. Attempts were made to rise to the challenge by using ingenious technology. The by then tried-and-tested technology for raising heavy objects was the waterwheel, powered by a conveniently adjacent fast-flowing stream. People had ways for raising water if a shaft could be driven. And they had ways to drive a shaft if there was a running stream.

Unfortunately, not all mines are located beside streams that flow year-round. Having animals walk around in circles as the driver incurs substantial expense. Over the ensuing centuries, this exact pressing need drove the invention of the steam engine, an engine that depended upon the first high-pressure pipework.

Figure 32 – A water-powered wheel pulls a hoist to raise water out of a mine (Agricola)

Well into Ottoman times, there was a continuity of earlier practice, and the odd sporadic spark of innovation in the domain of fluids. *Taqi al-Dín* (d. 1585) built a number of interesting machines in the tradition of that *Knowledge of Devices*. A water-powered scoop wheel served as the driver for a six-cylinder monobloc piston pump. The block rested directly in the water, and at the bottom of each cylinder was a non-return valve. Pipes led out from the bottom of the cylinders, brought together above the machine into a single delivery pipe. In a cliché to end clichés, he made a *steam-powered machine*, better than that of the ancient Greeks, but it was still only powerful enough to rotate a spit of doner kebab! [13]

Near Lugano, in the Italian part of Switzerland, *Agostino Ramelli* (d. 1610) published over 100 water-raising machines, including pumps, wells and fountains. But unlike his Ottoman predecessor, he would not be writing alone.

## Late medieval to early modern

Many manuscripts from those times are scattered in libraries and private collections, a corpus which is finally being painstakingly examined by academics. Even those that are housed in libraries (often colonial libraries) are mostly unavailable to share here due to murky copyrights. Some machines described therein have been reconstructed, offering glimpses into discoveries that formed the basis of what would follow. Other discoveries were forgotten, lost to time until scattered parts of a manuscript emerges.

A great deal of medieval know-how was lost and had to be rediscovered. Their legacy lives on in guilds all over the world, but the tricks of their trade weren't formalized into a corpus as early as mathematics, physics or chemistry. In fact, oftentimes they intentionally *concealed* their expertise, insisting that an apprentice must be overseen, approved and initiated by a master of their craft. There's probably a lesson there for us about engineering ethics: today's sterile training placing no emphasis whatsoever on the moral significance of our work, as if it were divorced from the realities of this world. Engineers are told to turn abstract concepts into facts on the ground, but excepting a select few sensitive industries, nobody at any point of our career cares to check that we are upstanding individuals! Although there are professional institutions, membership of one is *optional* in many places, and where membership is compulsory it doesn't tie engineers to a meaningful code.

Thankfully for our book, dissemination of knowledge became unstoppable. Slowly but surely, secrets could no longer be held so tightly bound. Open access inevitably brought with it new forums for technical discourse and critical peer-review. So the march of progress accelerated. The ideas developed so far, on motion, machines and materials, took fluids to exciting new heights.

As the world entered an unprecedented new period of enlightenment, no firm lines demarcated engineers from philosophers, or scientists from medics. There was ample room for anyone interested to turn their attention to the wonders of fluids. As they explored what may be perceived as mundane problems on flowing liquids and gases, they advanced in hitherto elusive disciplines like meteorology, and inadvertently invented kinetic theory.

# B4 – THE ADVENT OF SCIENCE AND ENGINEERING

## 1600's – The renaissance

### *Atmospheric pressure and the vacuum*

Based on careful observations with his telescope in Tuscany, *Galileo Galilei* (d. 1642) famously advocated heliocentrism in a book that overturned Ptolemy to side with Copernicus. For this the Roman inquisition declared him *"vehemently suspect of heresy"*. But that was not all he did. In 1594, the Venetian senate awarded Galileo a patent for a horse-powered device raising water, and soon thereafter he made a thermometer. When he wrote his ground-breaking work involving fluids, *Two New Sciences,* * he found nobody would print it in France, Germany or Poland. The church had secretly banned *any* book he ever writes! Eventually, *Lodewijk Elsevier*, Dutch namesake of today's publisher, was up to the task. On reaching Rome's bookstores, all fifty copies promptly sold out. Galileo's book recounts how a suction pump can lift water *"not a hair's breadth higher than 18 cubits"*, finding it notable, but as yet unable to explain why. It discusses a vacuum in contexts like his thought experiment dropping objects from the leaning tower of Pisa, but fluid *pressure* remained ambiguous.

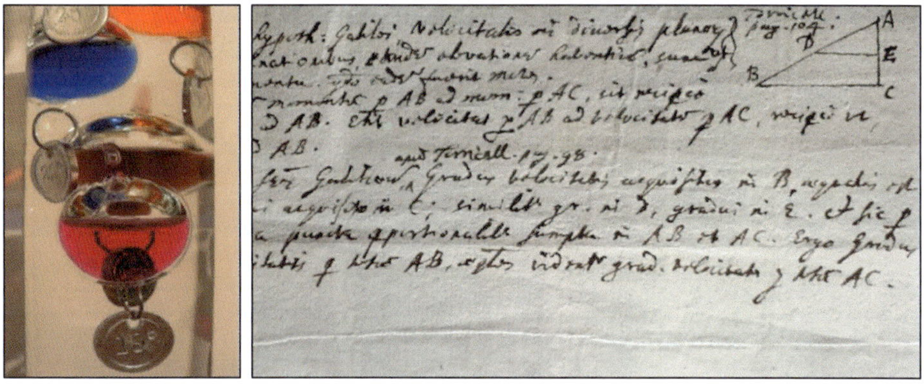

**Figure 33 – Galileo's thermometer relies on how temperature affects a liquid's density; beads of glass of different densities rise and fall whilst immersed within a column of liquid (left). Hobbes' notes on Galileo's theory of motion still found a vacuum absurd (right)**

The intellectual circle operating around Galileo outlived him. Shortly after his passing, his student, *Evangelista Torricelli* (d. 1647) said that we live immersed within a *"sea of air"*. This explains how *"winds are produced by differences in air temperature, and*

* *...a due nuove scienze;* the science of motion, and the science of materials.

*hence density, between two regions of the earth".* Air in the atmosphere exerts a pressure analogous to the pressure that water exerts on a submerged object. At sea level, this air pressure amounts to weigh the same as a column of water of a certain height: ten meters tall – those 18 cubits. When a simple suction pump – a rising piston – creates a vacuum inside a tube, the atmosphere no longer pushes on the water column below the piston but still pushes down on the surface of the water outside. This raises the water until its weight counterbalances the weight of the atmosphere. Torricelli ran an experiment lifting water and mercury through the same pump. Mercury being thirteen times heavier than water, the pump raised it one thirteenth (1/13) the height of water – 76 centimeters – a more convenient height than ten meters!

And so, in 1643, he filled a meter-long tube with mercury, sealing one end off, and set it vertically into a basin of the liquid metal. The column of mercury duly fell to about 76 centimeters. This momentarily produced a negative pressure – a partial vacuum – in the headspace above it. It is the first recorded incident in history where a vacuum could be maintained permanently, and the first time we see an instrument that can be really called a barometer: a pressure meter.

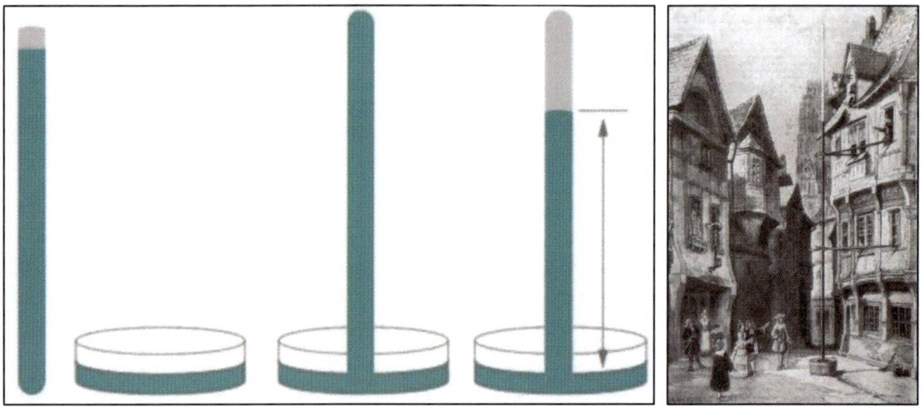

**Figure 34 –Torricelli's barometer measures atmospheric pressure (left) by gauging the drop in level of a liquid, using dense mercury rather than water to avoid unwieldy experiments (right)**

Torricelli discovered another important law. He found that water leaks through a hole in the wall of a container at a set rate, proportional to the square root of its depth ($H$) at an instance in time ($t$). A constant of proportionality ($k$) relates velocity ($v$) to the rate of change of depth over time ($\mathrm{d}H/\mathrm{d}t$):

$$\frac{\mathrm{d}H}{\mathrm{d}t} = -k \cdot \sqrt{H \cdot v}$$

Where velocity is, in turn, a function of depth:

$$v = v(H)$$

This finally formalized the long-exploited principle of the clepsydra, the water-powered clocks which were so finely tuned over the centuries since antiquity.

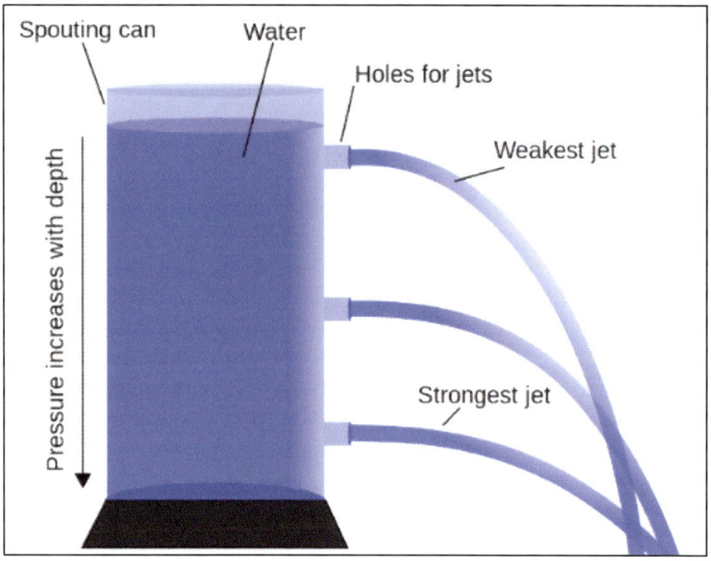


**Figure 35 – Pressure increases with depth in a static column of liquid, a fact long exploited for timekeeping for millennia – Torricelli quantified the resulting velocity**
**Matt Cook / Wikimedia Commons / CC BY-SA 4.0**

Meanwhile, in the German town of Magdeburg, *Otto von Guericke* (d. 1686) invented an air pump. As a politician, he demonstrated his invention to a public audience in an effort to impress. In the fashion of the era, and like many before him, he would forgo an explanation at times, leaving them in awe at his abilities.

His air pump pressurized a container by forcing air into it, or it sucked air out to depressurize it well below atmospheric pressure to create a partial vacuum. In 1654, invited to present before dignitaries at the Reichstag, he pumped the air out of a sphere, attached a team of horses to both sides and had them pull. They could not overcome the atmospheric pressure pushing in on the sphere.

In this way, his work was published for the first time, in an appendix, aged 55.

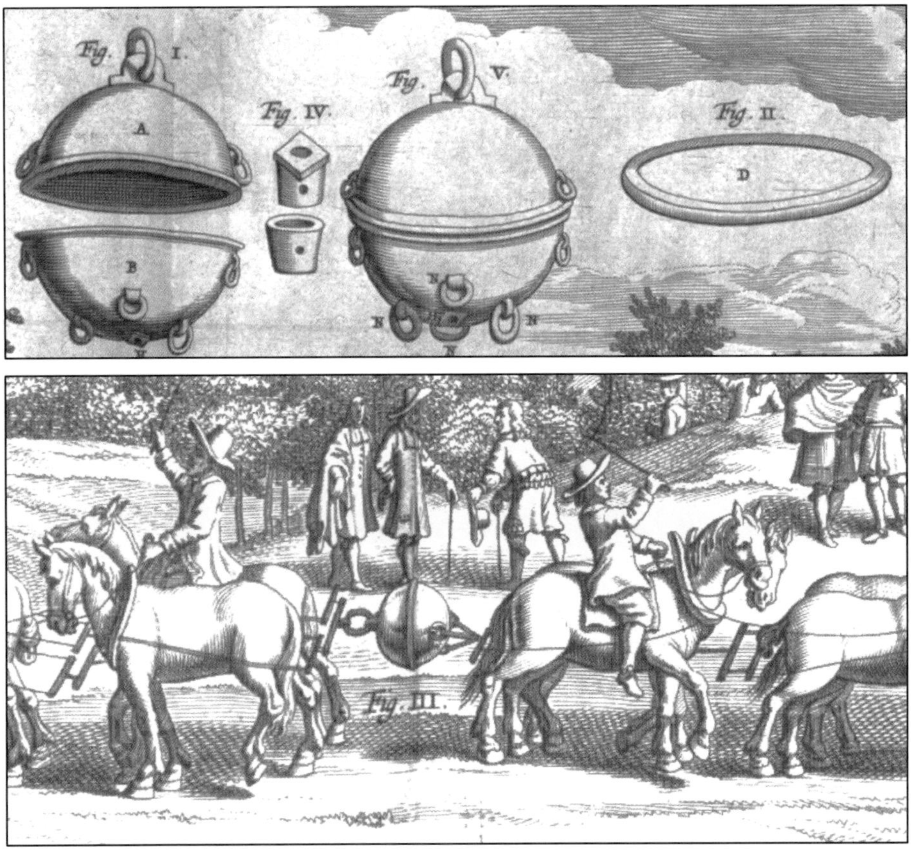


Figure 36 – **Air presses against an evacuated sphere so strongly that teams of horses can't pull it apart with their significant power. This was tried and illustrated by Otto von Guericke in 1657.**

By this time, a prodigy was turning 30. Educated by his father, a Rouen tax collector, *Blaise Pascal* (d. 1662) excelled in mathematics. As a teenager, he was (joint) first to invent a mechanical calculator. His name appears again and again in probability, economics, social science, and – of course – in fluids.

In 1647, Pascal rebutted Aristotle's followers who clung to the claim that nature abhors a vacuum. His results sparked heated disputes before being accepted. Torricelli had *"discretely sidestepped"* any mention of the church, which had decided by then that the concept of a vacuum was of theological importance, but Pascal – a theologian himself – wrote in defense of scientific method. This was a formative period. They weren't mere scientists, but *natural philosophers*.

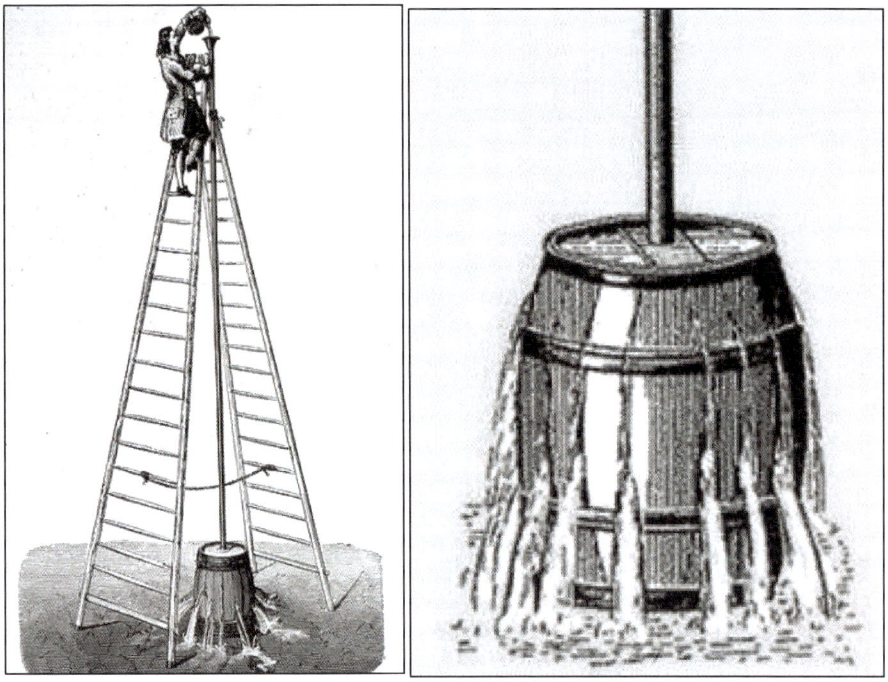

Figure 37 – The barrel buster, an experiment demonstrating hydrostatic pressure

Pascal's law dictates that pressure is transmitted everywhere within a liquid, undiminished in all directions. This important law underlies the principle of the hydraulic press. A famous experiment demonstrating it is the *barrel-buster*.[*] Insert a long, thin vertical tube into a barrel full of water. When a small amount of water is poured into the tube, the pressure beneath it bursts the barrel. Only *depth* ($\Delta z$) and density ($\rho$), not total weight, affect hydrostatic pressure ($\Delta P$):

$$\Delta P = \rho \cdot g \cdot \Delta z$$

It doesn't matter what diameter a pipe is, what shape it is, or what quantity of fluid lies above our point of interest. As a fluid sits stationary, it is subjected to hydrostatic pressure dependent only on the *height* of fluid sitting on top of it.

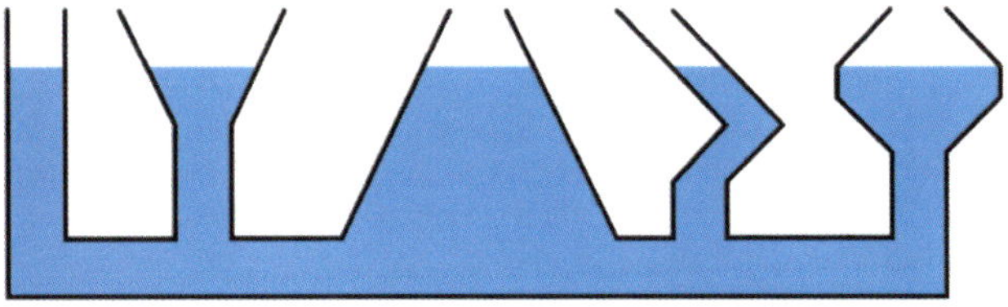

Figure 38 – The hydrostatic paradox: pressure at the base is independent of total mass above

* That Pascal *did* this experiment is possibly apocryphal, a later misattribution.

Pascal generalized Torricelli's work on fluids and made a second unambiguous prediction of the sea of air hypothesis, arguing that the mercury column of a barometer falls further at a higher elevation. He proved this by observing it fall slightly atop a 50-meter bell tower, and fall further atop a tall mountain peak. Changes in pressure at different elevations are exploited today by *altimeters*.

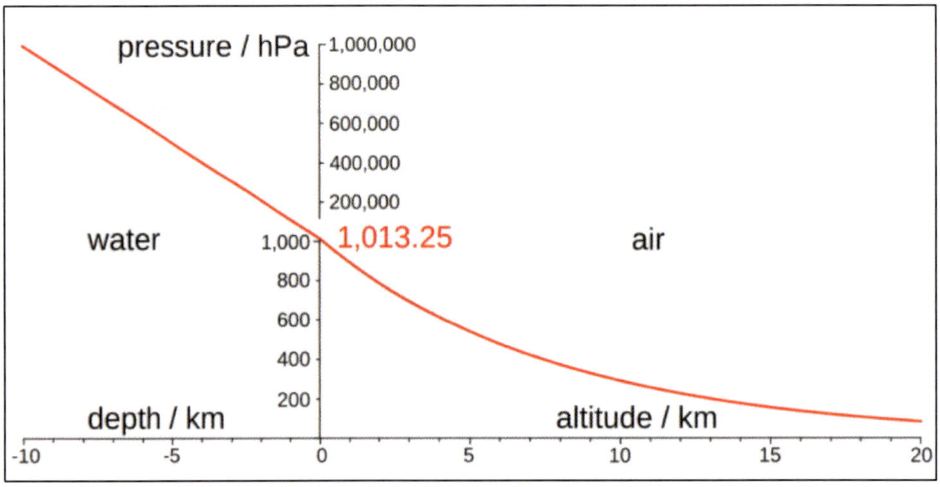


**Figure 39 – Altitude in the atmosphere is analogous to depth in water**
**Klaus-Dieter Keller / Wikimedia Commons / CC BY-SA 3.0**

People soon realized that over time, atmospheric pressure also fluctuates at the same location. They studied this phenomenon and started using it as a key parameter for the purpose of forecasting the weather: modern meteorology.

In honor of Pascal's work, the standard international system of units (SI-units) uses his name for a unit of pressure, the pressure exerted by the atmospheric air ($1\ \mathrm{atm}$) being a little over one hundred thousand Pascals ($\mathrm{Pa}$) in the metric system, which is a little under fifteen pounds per square inch ($\mathrm{psi}$) in imperial:

$$1\ \mathrm{atm} = 101\ 325\ \mathrm{Pa} = 14.696\ \mathrm{psi}$$

## *A gas law relating pressure and volume*

In 1661, on an ordinary hill in Lancashire, *Henry Power* (d. 1668) and his friend, *Richard Towneley* (d. 1707), conducted a set of experiments. Using Torricelli's barometer, they measured air pressure at different altitudes. They discovered a gas law: the volume occupied by a certain mass of gas varies inversely with its pressure, outlining it in a little-known book called *Experimental Philosophy*. That winter, Towneley went to London and discussed it with an aristocratic scientist called *Robert Boyle* (d. 1691). Boyle had improved on von Guericke's air pump, and so could experimentally verify this idea, and use his status to promote it. He cited *"Towneley's hypothesis"*, omitting any mention of Power. *

Boyle noticed that compressed air seems to be *"springy"*. The smaller the volume ($V$) it is squeezed into, the more it pushes back with its pressure ($P$). This holds true for all gases, at all the (near-atmospheric) pressures they could achieve at the time. The product of absolute pressure ($P$) and gas volume ($V$) is constant: so if pressure is, say, doubled, the gas will occupy half the volume.

$$P \cdot V = \text{constant}$$

In 1666, *Edme Mariotte* established an Academy of Sciences in Paris, as a rival for its London counterpart. In 1676, independently of Boyle, he discovered the same gas law, publishing it in his *"Discourse on the Nature of Air"*. Going further, he noted that the law holds only if there is no change in temperature, and coined the term *barometer*. In France, Boyle's law is called *Mariotte's Law*.

People realized, then, almost from the outset, that the value of this constant is affected by the type of gas, and by the temperature. Despite this, inserting temperature into this law would take another century. The reason is that at the time, nobody understood what temperature actually is, never mind measure it.

* Perhaps he feared confusion would ensue if it were called *"the Power law"*.

## *Newtonian physics*

All this was happening in a world very different to ours: a world before Newton. *Isaac Newton* (d. 1727) was more than a polymath – he left his mark all over science. In 1675, he famously said *"if I have seen further it is by standing on the shoulders of giants."* [*] Four of his ideas are important for us as pipeliners.

He laid down *conservation of momentum*, being mass ($m$) times velocity ($v$), known as Newton's Second Law. The momentum balance is a key idea for any engineer working with fluid flow. It implies a force ($F$) causes acceleration ($a$):

$$F = m \cdot a = \mathrm{d}(m \cdot v)/\mathrm{d}t$$

In 1686, he was the first to calculate the speed of sound ($c$) in air, by timing how long it takes an echo to traverse a colonnade. He was off by 15%, having no way to measure a third of a second. Short time intervals are always used in any experiment or simulator to analyze effects relating to sound speed in fluids.

$$c \approx 340\ \mathrm{m/s} \text{ in air at ambient conditions}$$

He introduced the principles of viscosity, and what we call a Newtonian fluid. His observations led him to deduce a liquid's viscosity ($\mu$) is constant at a set temperature, relating an applied shear stress ($\tau$) to resulting shear rate ($\mathrm{d}\gamma/\mathrm{d}t$):

$$\tau_{\mathrm{fluid}} = \mu \cdot \frac{\mathrm{d}\gamma}{\mathrm{d}t}$$

In the early 1700's, he formulated an empirical law now called Newton's Law of Cooling. "*The rate of heat loss ($Q$) of a body is directly proportional to the difference in the temperatures ($\Delta T$) between the body and its surroundings*":

$$Q = U \cdot A \cdot \Delta T$$

A big part of this book concerns momentum, sound speed, viscosity, and heat.

---

[*] Living in an age when discerning reason from superstition was a muddy task, he was arguably as much *"the last magician"* as the first in a new age of reason.

## *Gunpowder pumps*

In 1661, English inventor *Samuel Morland* (d. 1695) sought a way to raise water to the royal residence at Windsor Castle. His idea was to explode gunpowder in a pipe, creating a vacuum, sucking water up. He patented the idea, but never realized it. Mad as it is, gunpowder pumps changed the world.

In 1678, French inventor *Jean de Hautefeuille* (d. 1724) faced the challenge of raising water to the royal residence at the Palace of Versailles. He presented two ideas, both powered by gunpowder. One was to use Morland's vacuum. The other was to use a U-shaped tube with water in one side and air in the other. That way, when gunpowder was lit in the air-filled side, its pressure *rise* would indirectly create a suction effect at the other side, pulling the water up.

All of those early designs strived to directly suck water up, with no moving parts and no mechanical intermediaries. One fluid simply sucked up the other.

*Christiaan Huygens* (d. 1695), an able Dutch polymath, picked up that work. In 1680 he implemented his design for a gunpowder *engine*. Gunpowder was inserted into a vertical tube containing a piston. It was lit through a small hole at the base, like a cannon. The expanding gases would drive the piston up the tube until it reached a point near the top. Here, the piston uncovered holes in the tube that allowed any remaining hot gases to escape. The weight of the piston and the vacuum formed by the cooling gases in the now-closed cylinder drew the piston back into the tube, doing work by lifting a mass. A prototype was built, using a cannon as the cylinder. Two years later, he presented a paper on his invention: *"A New Motive Power by Means of Gunpowder and Air"*. His ideas on conservation of energy would inspire generations after him.

By 1682, there are reports Huygens had demonstrated that a dram (1.7 grams) of gunpowder, in a cylinder eight feet high and eighteen inches in diameter, could raise eight boys (holding onto the end of a rope) high into the air. Those reports are disputed, as attempts at modern recreations struggled to seal the piston within the cylinder. The science of materials – specifically of *fittings* – had to catch up with the science of motion. But his *"engine"* was revolutionary. Within a couple of decades, development efforts had turned to steam. Morland, the very first figure in our story of engines, had calculated very early on that the volume of steam was approximately two thousand times that of water. That immense difference in volume occupied by a liquid and its vapor would surely create an almighty vacuum, a real concern for today's pipeliners who strive to avoid inadvertently create a bubble of vapor within their hydrocarbon liquid, but an asset for those early inventors looking to generate power with an engine.

A star student of Huygens, Frenchman *Denis Papin* (d. 1713), ran experiments on the vacuum, and the machines used for making it, publishing in 1674. He conceived many devices using air pressure, including fountains and pumps, and built a clepsydra – which Huygens reviewed (presumably before inventing the pendulum clock). In 1679, Papin invented the steam digester, a forerunner of pressure cookers, exploiting steam's incredible *latent heat of vaporization*.

Not long after inventing that pressure vessel, he stopped it exploding. He achieved this by a principle we now call *pressure relief*, using a well-calibrated *safety valve*. He was mesmerized by how steam made this valve rhythmically rise and fall, and managed to make piston raise weights by steam in 1690. However, he could not take the idea any further. He fell into destitute poverty, and died in obscurity. But he had planted the seed. Before long, everyone realized their future would be steam powered.

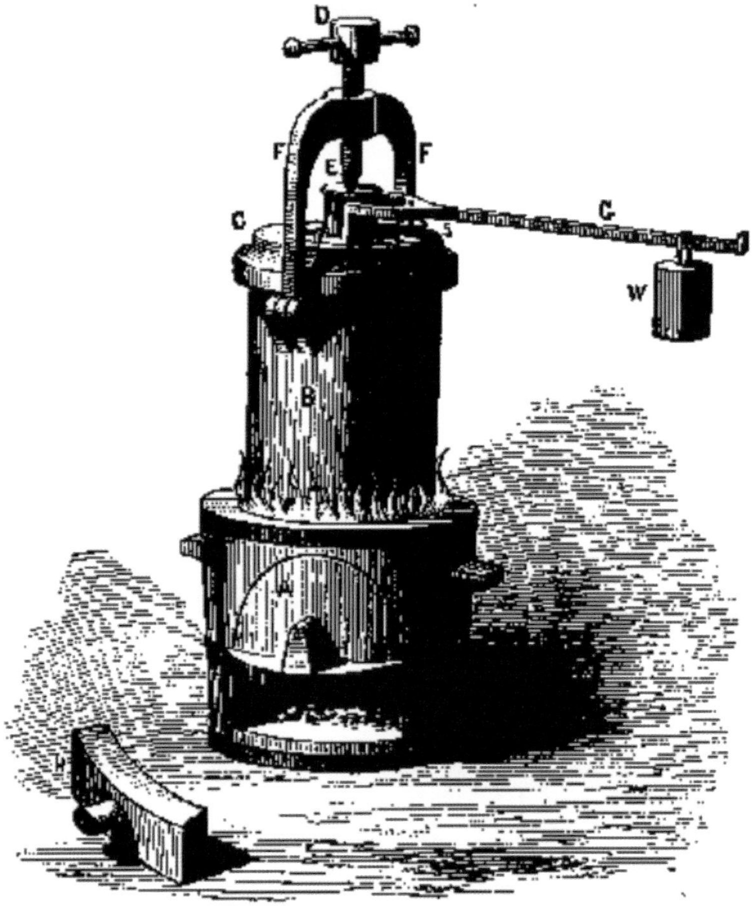

**Figure 40 – Papin's 1679 steam digester had a relief valve that made the pressure vessel safe**

# 1700's – The enlightenment

## *Early coal, steam and electricity*

In 1702, *Thomas Savery* (d. 1715) invented *"The Miners Friend; or, An Engine to Raise Water by Fire"*. With no pistons, powered only by the pressure of the atmosphere against a vacuum, the way that this pump worked was quite mad.

First, it raised steam in a boiler. Then it admitted the steam to a working vessel, allowing it to blow out through a downpipe into the water that is to be raised. When the system became hot it filled with steam. Then a tap between the boiler and the working vessel was shut. The outside of the vessel was then cooled, condensing the steam inside it, creating a partial vacuum. atmospheric pressure pushed water up the downpipe until the vessel is full. At this point the tap below the vessel was closed, and the tap between it and the up-pipe opened, and more steam was admitted from the boiler. As the steam pressure built up, it forced the water from the vessel up the up-pipe and out of the mine.

Needless to say, using condensing steam to invoke a vacuum that sucks water up wasn't ideal. Savery's steam-powered pump had some serious drawbacks. Heat was wasted warming up the water that was being pumped, because it was being introduced to the same vessel as the steam resided! Then, in the second stage of the process, when high-pressure steam was admitted, the soldered joints were barely capable of withstanding those pressures and often gave way. Finally, water was not pushed up by steam pressure but sucked up against a condensed-steam vacuum, with atmospheric pressure doing all the work. This restricted the pump to be no more than 9 meters above water level. To drain a deep mine, engines were placed underground, one above the next!

Those fits and starts helped Papin, inventor of the safety valve in the 1600's, to develop even more innovations with the help of *Gottfried Leibniz* (d. 1716), a polymath who was, like most polymaths it seems, working to design his own pumps for draining mines. Papin improved upon Savery's machine by using the pressure of steam rather than the pressure of the atmosphere, publishing a new pump design in 1707. His papers were put before the Royal Society without any acknowledgement or paying him his due credit, a bitter grievance.

**Figure 41 – A diagram (left) and a reconstruction (right) of Papin's 1707 steam engine. It had high-pressure pipework exploiting the energy of pressurized steam to raise water. Like all early steam engines, it was in fact a pump to move fluids rather than an engine.**


In 1712, *Thomas Newcomen* (d. 1729), an ironmonger specialized in the design, manufacture and sale of tools for the mining industry, took forward Papin's experiment and made it workable. He installed his first working steam-powered pump at a mine in Staffordshire, England. Working at 12 strokes a minute, it could raise water from a depth of 156 feet. Its efficiency was still only 1%: these contraptions remained too unwieldy and too inefficient for wider application, fit only for pumping water out of mines. But they were good enough in the context of coal mining to prove a viable replacement for animal power, especially as they could work day and night, and were powered by the same coal they were helping to extract. By the time he died, over a hundred of his steam engines were providing unprecedented power on demand.

During the course of the 1700's, the precursors to the industrial revolution were mills processing wool, then mills processing imported cotton. First powered by water wheels, then powered by the emerging steam engines, their mechanical looms used punchcards to direct the weaving of patterns, simple algorithms that were the predecessors of computer programming. Around the mill towns there grew countless industries dying, tanning, and manufacturing all kinds of wares. The coal mines became busier than ever. Goods were now being manufactured at unprecedented rates across Europe, presenting a need to transport coal and feedstocks to the factories, and products to their customers.

New canals were built, giving rise to a brave new world of international trade. Raw materials had always been shipped across vast seas and oceans, and converted to finished products to be traded on again. But never in recorded history had it been an enterprise on this scale. Global markets emerged, and the forces of supply and demand drove the development of new products, and the processes to make them. The copious supplies of coal gave people ideas.

The first question was: is it necessary to use so much of it? At the University of Glasgow, instrument maker *James Watt* (d. 1819) conducted experiments on steam power. He realized that a steam engine that better conserved energy would be revolutionary. To advance this research, he founded a company near Staffordshire in partnership with a manufacturer, *Matthey Boulton*. Their firm, *"Boulton & Watt"*, advanced steam power, creating the world's first versatile steam engine. Instead of heating and cooling the same vessel in successive inefficient steps, they built a cold condenser separately, placing it away from a hot boiler. They charged a set annual fee equal to a third of the value of coal saved over a Newcomen engine. This was quite the value proposition. Their engines rose to fame, finding widespread use driving all kinds of machines. But Watt's happiness was not to last.

Indeed, could the coal itself be processed into more useful products?

When Watt tragically lost his daughter to consumption in 1794, he resolved *"that when an evil is irreparable, the best consolation is to turn the mind to any other subject that can occupy it for the moment"*, experimenting to make gases that may avert future suffering. In doing so, he did generate an anesthetic, though societal barriers stood in the way of its uptake for decades thereafter. His gasification experiments had further ramifications: he had thereby invented a process to manufacture coal gas. One of his employees, *William Murdoch* (d. 1839), worked on scaling up the process and using it for gaslighting. At the same time, in France, *Philippe Lebon* (d. 1804) was doing much the same. [42]

## *The temperature of a gas*

All gases were *"new"* at the time. *Antoine Lavoisier* (d. 1794) was a chemist and a humanitarian who involved himself in all kinds of philanthropy, including an aqueduct, and street lighting. He focused his attention on water quality, air quality and the properties of these new gases. Investigating Henry Cavendish's *inflammable air*, he preferred to rename it *hydrogen* (the water-former), and quantitatively proved the ancient Greeks – and anyone who still believed them – to be wrong again: air is *not* an element. Indeed, as *John Dalton* (d. 1844) soon established, a gas mixture exerts an overall pressure ($P$) that is the sum of its components partial pressures ($P_i$), according to their molar fractions ($X_i$):

$$P_i = X_i \cdot P$$

Another question was temperature. Lavoisier rightly noted that the prevalent theory of *phlogiston* – heat as a material – was *"inconsistent with experiments"*, by carefully weighing reactants and products on delicate scales, though he couldn't quite phrase what heat actually *was* yet. His experiments happily led to the end of phlogiston, an embarrassing chapter in science, but sadly he didn't live to see it. In the French Revolution, he was guillotined, aged only fifty.

*Guillaume Amontons* (d. 1705) improved the barometer and thermometer, and though the latter was still not good enough to reach a quantitative conclusion, he investigated the effect of temperature on gas pressure, finding pressure to reduce by a third between the temperature of *"cold"* and boiling water. He realized that eventually, at some exceedingly low temperature, absolute zero, the pressure would disappear altogether. Soon, *Daniel Fahrenheit* (d. 1736) made the first accurate, practical thermometer. He defined a temperature scale of his own, as did *Anders Celsius* (d. 1744), allowing standard measurements.

In the 1780's, *Jacques Charles* established that the ratio of gas volume ($V$) to its temperature ($T$) is constant, provided the pressure ($P$) remains the same:

$$\frac{V}{T} = \text{constant}$$

## *The flow rate of a liquid*

*Henri Pitot* (d. 1771) rose to fame as the engineer of several aqueducts. One was an extension of the *Pont du Gard* seen earlier, literally picking up where the Romans left off. Tasked with determining the flow rate of the river Seine, in 1732 he invented a measurement device to work by an interesting principle.

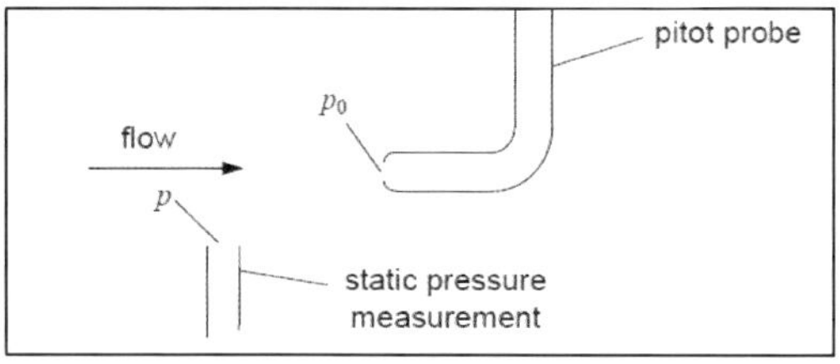


Figure 42 – A Pitot probe is a tube whose inlet faces the oncoming flow. The height a liquid rises within its column can be used to determine the velocity of the fluid flowing at its inlet.

A tube facing sideways is a manometer or a piezometer measuring the *static* pressure. A *"Pitot tube",* faces the direction of fluid flow, measuring a higher *total* pressure. The difference is a *dynamic* pressure giving us the flow velocity, in a manner described by Bernoulli's Principle or more generally by the law of conservation of energy, as we will cover in a chapter of the main book.

In Switzerland, *Leonhard Euler* (d. 1783), combined an interest in fluids with mathematical genius. * His approach observes a *control volume* bounded by a *control surface*, through which fluid enters and departs as it flows. This lies at the heart of many modern computational pipeline models to this day. In 1757 he published equations governing inviscid flow – called the Euler equations.

Meanwhile, in Turin, a 17-year-old *Giuseppe Luigi Lagrangia* (d. 1813) saw a paper by *Edmond Halley,* and became a self-taught mathematician teaching geometry at the Royal Artillery School within a year. He was invited to succeed Euler as a director at the Berlin institute, then moved to Paris by which time he went by the supposedly more fashionable French version of his name: *Joseph-Louis Lagrange*. He built on the earlier work by Newton formulating a basis for mathematical physics to develop dramatically. To study the flow of fluids, Lagrange's approach follows a clump of fluid material while it moves along, a perspective called a *control mass* (or, confusingly, a *"system"*).

* The base of the natural logarithm is called Euler's number and represented *"e"* in his honour. Confusingly, there are also several other numbers defined in his honour. An unrelated dimensionless *"Euler's number"* and represented *"Eu"* is used for fluid flow.

One can intuitively reach the exact same conclusions and results by either of the two perspectives; an Eulerian control volume or a Lagrangian control mass.

Switzerland was an intellectual hub in this period. Hailing from a long line of Dutch mathematicians residing in Basel, * *Daniel Bernoulli* (d. 1782) published his seminal work, HYDRODYNAMICA. In Latin, he applied conservation of energy, received from what Huygens had written in Holland. He described how water flows through a tube, and from a hole in a container. He explained dynamic pressure, and how a loss of kinetic energy affects flow. He introduced notions of *work* and *efficiency* to hydraulic machines. He discovered the kinetic theory of gases, demonstrating that the pressure of flowing air (a bulk phenomenon) is proportional to the kinetic energy of each particle of air (at a molecular level).

> principia inter ſe admodum ſimilia; ſcis enim quantitatem *virium vivarum* corpori inſitarum deſignari per $\frac{1}{2}mvv$ et eſſe $\frac{1}{2}mvv = \int p\,dx$ (deſignato per $dx$ elemento ſpatii, quod tempuſculo $dt$ percurritur), id eſt, =

**Figure 43 – In 1741, Bernoulli published the multiplier of one half in defining kinetic energy**

Working with Euler, Bernoulli sought a way to relate blood pressure to the speed it flows. They experimented by puncturing a pipe's wall with an open-ended straw, and found that the fluid rose up to a height related to its pressure. Soon, physicians all over Europe started sticking point-ended glass tubes directly into patients' arteries to measure their blood pressure. 170 years later, an Italian doctor discovered a less painful method which remains in use today.

Taking his discoveries further, Bernoulli revisited the question of conservation of energy. It was known that a moving body exchanges its kinetic energy for potential energy when it gains height. He realized that in a similar way, a moving fluid exchanges its kinetic energy for *pressure*. Bernoulli's Principle – that a faster-flowing fluid actually has *less* pressure – is embodied within the famous Bernoulli's Equation, the equation describing the energy of a fluid flow.

* A family that developed jealous feuds over who wrote what first about fluids.

## *The metric system*

The imperial system of currency, weights and measures was popularized in Europe by *Charlemagne* (700's), and is the successor to Roman practice. Decimal notation had appeared sporadically throughout the middle ages in various guises and with various notations. Decimals came to be widely used in Arabic texts, and before that in India and China, but it was transmission to Europe that paved the way for decimalization. In 1202 Fibonacci introduced Europe to Arabic numerals in LIBER ABACUS (his *Book of Calculation*). In 1585 Simon Stevin published a pamphlet (*The Tenth*) in Dutch, then in French. It introduced Europe to decimals, predicting decimalization of coinage, weights and measures to be merely a question of time. That time came 200 years later.

When pendulum clocks were invented in the late 1600's, there were proposals to use a decimal system for subdividing a day into hours and minutes, but the long established practice using 24 hours and 60 minutes was too entrenched.

In 1783, frustrated by difficulties communicating with German scientists, Watt called for the creation of a global decimal measurement system, proposing the density of water to link length and mass. In 1788 Lavoisier commissioned a set of brass cylinders for his experiments, one weighing a French pound and nine more weighing decimal subdivisions thereof. The French proposed to define a meter as the period at which a pendulum beats over one second. The idea was defeated in both the British parliament and American congress in 1790.

But the time had come. Slowly but surely, decimalization took hold over the subsequent centuries, with anyone using the new system never looking back. * Unlike traditional units, metric units proved to be easy, sensible and consistent.

---

* Today, only three countries don't adopt the metric system: Myanmar, Liberia and the United States of America. The influence of the latter's companies on engineering sectors results in some peculiar combinations of metric *"SI units"* and traditional *"imperial units"*, varying by country, by company and by team. Conversion error famously brought down a NASA spacecraft, and has been the direct cause of major incidents and accidents in industries such as aviation. There are other countries that use imperial units alongside the metric system in daily life, such as the United Kingdom, but they do not do so in engineering.

# 1800's – The industrial revolution

## *Steam engines, boilers and ships*

Steam engines were excellent in stationary application, but were initially too heavy to shift their own weight as a locomotive. This presented an interesting new problem. Not long after the canals, railways were built. Was there a better way to move loads along them than by horse? *George Medhurst* (d. 1827) proposed a solution: to propel the trains, build evacuated pipelines alongside the tracks. These were atmosphe*ric railways*: a neat solution for circumventing cast iron's weakness in tension, because the air inside this pipeline is at a partial vacuum, so all the iron had to do was resist the inward compression of the atmosphere. The differential pressure between the atmospheric air and the vacuum in the pipeline was enough to drive the trains along the railway.

**Figure 44 – A cast-iron pipeline carried steam for an *"atmospheric railway"* in Devon, England**
**Chowells / Wikimedia Commons / CC BY-SA 3.0**

As appealing as the idea was, all of the 19th-century atmospheric railways were short-lived, being replaced by conventional steam locomotives after a few years due to assorted technical difficulties. Railroad charts were used to schedule the logistics of movements. Busy sections of track were replaced often, until in the mid-1850's, Bessemer invented his economical steelmaking process. Steel vessels could be built to withstand incredible pressure.

*"Although there are innumerable accounts of the history of the steam engine, relatively little has been written about the development of the pipes and boilers on which it depended and which, in reality, presented more difficult problems than the actual mechanism"* [14]

The industrial revolution was about *processes*. Grand feats of civil engineers. Unprecedented products made by chemical engineers. And, of course, steam engines – the literal powerhouse – designed by the mechanical engineers.

The already influential mining industry became even more important, as coal became power. So much soil was shifted in unearthing coal from the Ruhr valley since 1850 that the subsidence caused by removing so much material caused the entire region to sink by as much as 20 meters in places, altering the flow of water and the courses of streams. Today, sixty pumping stations are kept in constant operation to drain water from that low-lying ground, their pumps are never allowed to fail or else whole swathes of that part of Germany would immediately flood.

Making steam engines that aren't comically colossal, impossibly heavy or ridiculously inefficient is all about *working pressures*. In the 1820's, steam engines were achieving pressures of a mere ten pounds per square inch: 10 psi – a mere two thirds of atmospheric pressure. [*] In 1850, steam engines were still at 20 psi. Pressure boilers were pushed ever higher by daring risktakers: Mississippi steamboats racing over thousands of miles, their designers willing to do anything to win the race. In two years alone, twenty-seven of these vessels were lost to boiler explosions. [†] By 1900, steam engines reached 200 psi, improving fuel consumption tenfold. It was not steam that *"drove sailing ships from the high seas"*, it was *high-pressure* steam. [14]

*William Fairbairn* (d. 1874) was one of the first engineers to conduct systematic investigations of failures of structures, including boiler explosions. His research into metal fatigue quantified the practical yield strength of a pressure vessel. And through experiment he determined the hoop stress in its wall to be twice the longitudinal stress, explaining why when a cylindrical vessel fails, it usually fractures along its length owing to the high hoop stress in the wall. This meant engineers could build boilers, pipes and vessels that work at higher pressures.

[*] In 1835 Galileo's works were finally taken off the list of forbidden books.

[†] Plus 83 destroyed by fire, 88 ran into sunken trees, and 70 by "other causes"!

The broad remit of ironsmiths carried on as late as 1894, their medieval skill set still proving indispensable. It culminated in revolutionary creations like the *Turbinia,* the first turbine-powered steamboat, made not by shipbuilders but by an ironworking firm that had no prior experience in the matter. That ship moved so fast that its propeller – referred to as a *screw* by naval personnel – had to be redesigned. It needed to overcome a new phenomenon: *cavitation*. That same design challenge still applies to pumps in modern pipelines, and so is an equation presented by engineer *Louis Charles Antoine* to address this in 1888.

Steam found innumerable applications beyond steam engines. In New York City, *Wallace Andrews* built a steam pipeline system in 1881, to distribute energy from his plant's boilers to hundreds of customers. In 1888, a blizzard destroyed the city's above-ground wiring. When they re-laid it underground, they lay steam pipes alongside the electric cables. Every building was supplied by utility steam along with electricity. Whenever work was done carelessly near these lines, a phenomenon occurred known as *steam hammer*, causing steam to erupt out of the ground on occasion. [15] Thankfully improvements were implemented, requiring permits to work in the vicinity of the pipes!

## *Waterworks, water mains and sewers*

Figure 45 – Crossness hosts possibly the world's most beautiful sewage pumping station. The engineer overseeing these works commissioned 2 000 kilometers of pipelines across London, intentionally oversizing them so they stay future-proof.
Peter Scrimshaw / Wikimedia Commons / CC BY-SA 4.0

As cities grew during the industrial revolution, new infrastructure was built. So wretched was the river Thames during the great stink of 1858, that Parliament could not breathe. Great *main sewers* were built, as well as new pipelines to carry fresh water below London (700 km by 1870) and Paris (600 km by 1878). The challenges of pipelining – pumping stations, pipe sizing, flow rates, and so forth – suddenly appear during this period, as engineers of the age struggled to apply the principles of fluid flow in practice.

It is interesting that the event prompting that system was a great *stink*. Established opinion was that *bad air* was to blame for diseases, not bad water. In London, a cholera outbreak broke out (again) in 1854 and killed 616 people in just two weeks. Determined to get to the bottom of it, a bold physician – *John Snow* – meticulously went house to house, drawing up a map that correlated cases very clearly against whether a place was being supplied by water from the *Broad Street Pump*. He found that pump to be contaminated. His theory went against common knowledge, but through perseverance, meticulousness and tenacity, his overwhelming evidence meant that none could argue against water pollution being the culprit. Everyone was wrong: water quality matters!

## *Gasworks, gaslighting and gas-fired engines*

The *"thermolamp"* invented by Lebon at the turn of the century was great, but it didn't find immediate success. It took two decades of persistent marketing to convince investors and governments that building gas infrastructure is worth the trouble and expense. Part of the challenge was explaining what it even was. Members of Parliament and of the Royal Society assumed the pipes would be hot because they carried an invisible substance that was meant to be lit on fire. Lebon himself expended every effort to carefully explain it all to the sceptics. [42]

> *"This ethereal substance can travel, in a cold state, along a pipe only an inch square within the thickness of walls... you can control it, order it to appear or disappear"*

A German chemist painstakingly testified before parliament: [42]

> that its flame *"may be increased or diminished by admitting more or less of the gas"*, that it is *"contained in the pipe... by turning the cock you let it out"*.

But unconvinced, the famous Humphry Davy scoffed: [42]

> *"you would have to fill St. Peter's dome to get as much gas as you need, and then it would explode"*.

Such was the resistance to gas technology. Yet despite this, within decades, every major city in Europe had a gas network. Gas lighting heralded the dawn of a new era for dark streets after nightfall, so dangerous that in Manchester, gas infrastructure was initiated by the police! The first gas pipes were laid by gunsmiths welding musket-barrels end to end, and by 1819, 288 miles of pipes and meters supplied 50 000 burners across London. [16] Flammable towns gas was manufactured from coal at gasworks, in large retorts scaling up the principle of the alembic from Watt's laboratory.

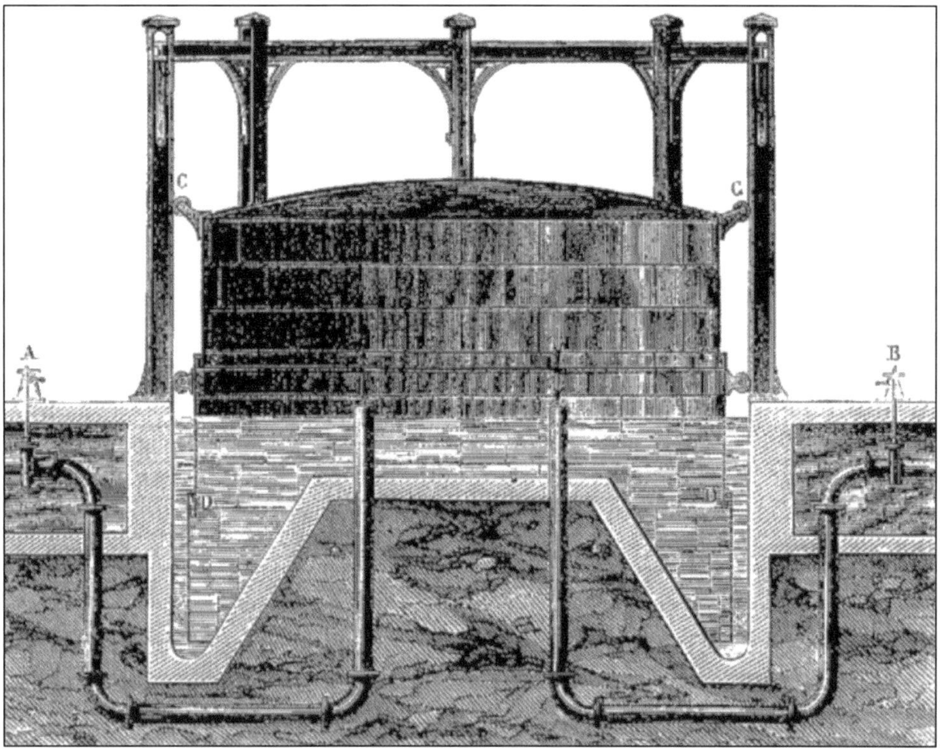


Figure 46 – A telescopic gasometer, illustrated in 1867 (*Les Merveilles de la Science*)

Huge vessels known as gasholders or gasometers balanced fluctuations in supply and demand by storing surplus gas during periods of little demand and using them as buffer supplies during peak periods. As the gas is stored at or near atmospheric pressure, the entire floating roof of the gas holder rises and falls depending on the quantity being stored at any given instance. A governor housed in a small adjacent room controlled the flow out of the holder by adjusting a control valve acting as a pressure regulator.

> Each of these vessels holds 50 000 cubic meters of gas at atmospheric pressure. Today, linepack holds that much gas within a 50 meter stretch of high-pressure pipeline. If it's at 100 barg pressure, what diameter pipe is needed?

Coal was soon shipped by railroad to upstate New York, and coal gas pipelined alongside the newly built Erie canal, infrastructure linking the Great Lakes to the Atlantic Ocean. This infrastructure gave the *empire state* its new megacity.

Throughout the late 1800's, processes were developed to manufacture gas from coal (in Europe) or wood (in the United States) with more desirable properties. Passing steam over hot coal generates a proportion of hydrogen so that it burns more effectively. Washing the gas removes its odor. The processes and terminology varied; what mattered was the resulting properties.

The gas companies were all about light until electricity arrived on the scene. A naked flame of gas – or several flames – were intended only for lighting, their heat simply going to waste. In 1885, an Austrian chemist invented the gas mantle, a fabric gauze coated in chemicals that illuminated when heated. But the first electric bulbs had already been patented in 1879 by Joseph Swan and Thomas Edison (independently and almost simultaneously). So why gaslight? Electric arcs were the most popular type of electric lighting, but required high voltages incompatible with Edison's low-voltage incandescent system. And although electric arcs were the preferable of the two, their tendency to flicker and hiss remained a major problem. A coal gas flame might smell, but for a while gaslighting remained the preferred option for a great many customers.

In 1895, *Hertha Ayrton* wrote a series of articles, explaining these hissing phenomena arose as oxygen came into contact with the carbon rods. A close friend of *Marie Curie*, and an active aid to those imprisoned and killed in the struggle for women's suffrage, her biography reads like a long series of firsts. Her activism brought her into contact with the founder of Girton College, the first college to admit women in Cambridge, [*] who financially supported her education and career. In 1899, she was the first woman to read her own paper before the Institution of Electrical Engineers, and was elected its first female member shortly thereafter. [†] She petitioned to present a paper before the Royal Society but was rejected for being a woman. Electric arcs would prove to be important again in history, as welding enabled high-pressure pipelines.

Until very recently lightbulbs essentially dumped heat while giving a little light, and for a while they remained inferior to the illuminating gas mantle. In due course, lightbulbs got better, displacing gas lighting, then becoming so good that manufacturers faced disappointing sales as bulbs didn't need replacing!

In 1859, Étienne Lenoir built the first successful *"explosion engine"*. [‡] He threw away the boiler of a steam engine and modified it to instead run on the heat generated from the ignition of *"illuminating gas"* now being pipelined all over Paris. Despite the violent explosions damaging pistons and valves, rendering his invention less efficient than the

---

[*] The University of Cambridge only accepted this college (and women) in 1948.

[†] Britain's Institution of Electrical Engineers next admitted a woman in 1958.

[‡] The original – arguably more accurate – name for a combustion engine.

original steam engine, his gas engines were the first to be put into serial production, and he sold around seven hundred.

His work inspired another budding entrepreneur: a German wholesale grocery salesman named Nicholas Otto. After a great deal of determination, he came up with a more efficient design in 1864, selling many thousands of his engines. But they had hit a technical dead end: it produced only 3 horsepower yet could only be installed where two floors of headroom were available. He solved this problem with yet another innovation in 1876, a four-stroke compressed charge cycle for which he remains the namesake of modern combustion cycles today:

- A downward intake stroke admits fuel to a combustion chamber
- An upward compression stroke forces a piston to reduce the volume
- A power stroke ignites the fuel-air mixture using a flame (later a spark)
- An exhaust stroke releases products of combustion from the chamber

The efficiency of this cycle became a preoccupation for mechanical engineers and physicists, who had to retrospectively theorize their way through what was happening in these machines. This became a cornerstone of thermodynamics.

## *Gas laws and thermodynamics*

In the 1800's, *Joseph Gay-Lussac* (d. 1850) established that pressure ($P$) is proportional to temperature ($T$) at constant volume. Gas pressure, volume and temperature seemed intertwined, but it would still take decades to formalize.

In 1811, *Amedeo Avogadro* (d. 1856) discovered that the same quantity of *any* gas held at atmospheric pressure and room temperature occupies the same volume: just over 22 liters. Incredibly, this holds true for *any* pressure and temperature: the same quantity of any gas occupies the same volume. This breakthrough observation helped him to define the concept of *quantity* in terms of *number of molecules*.* As the number of molecules is mindboggling, we speak in terms of a multiplier known as *Avogadro's number* ($N_A$) in his honor:

$$N_A = 6.02 \times 10^{23}$$
$$= 602\ 000 \text{ million million million}$$

* The concepts of atoms and molecules remained controversial well past 1900.

This is called a *mole*. There are fewer stars in the observable universe than atoms in a mole of fluid. He found the volume ($V$) of gas in a container divided by the number of moles ($N$) is constant, if pressure and temperature are equal:

$$\frac{V}{N} = \text{constant}$$

In 1857, *Benoît Clapeyron* combined the other three earlier gas laws to say:

$$\frac{P_1 \cdot V_1}{T_1} = \frac{P_2 \cdot V_2}{T_2}$$

If we draw a three-dimensional *P-v-T* surface with lines of constant pressure (isobars), lines of constant temperature and lines of constant specific volume (constant density - isochores), we see the earlier laws fall out of projecting this relation in two dimensions.

The final *ideal gas law* therefore involved a new *universal gas constant* ($R$):

$$P \cdot V = N \cdot R \cdot T$$

Pipeliners express this using *density* ($\rho$) and the *molar mass* ($m_r$) of the gas:

$$\rho = \frac{m_{\mathrm{r}} \cdot P}{R \cdot T}$$

This is called an equation of state. It holds true in the *"ideal conditions"* of those early experiments: near-Atmospheric pressures and near-room temperatures.

The pioneers of the industrial revolution in the early 1800's were not concerned with *why* anything happens, but with *how* it can be made to work in practice. A backlog of unaddressed questions was building up. To tackle them, scientists had to revise long-standing ideas on energy and heat, discard caloric theory, define temperature and lay the foundation for a new science: thermodynamics. It is so important that we ought to touch on their work, though few attempt this.

We prefer to focus on figures who worked with fluids and hydraulics, rather than those who turned their attention to thermodynamics, for two reasons. Firstly, it's so difficult that even its history is rather convoluted. Secondly, many leading thermodynamicists met a tragic end, often by their own hand. Narrating their personal history therefore runs a real risk of dampening everyone's mood. Nonetheless, it is worth being familiar with the rudimentary origins of the ideas we shall be using later in this book, if only as an ice-breaker for a reader who hasn't come across them before. People only figured all this out quite recently!

Thermodynamics dictates that whenever energy is converted from one form to another, we lose some of it to inefficiency. This inherently incurs an expense. Yet fossil fuels turned out to be so cheap that despite dreadful well-to-wheel efficiencies even today, cars still rely on them over other sources of energy. It is easy to get sucked into theorizing and lose oneself in trying to wrap our head around thermodynamics, thereby losing sight of the bigger picture. This is why any good course on the subject is grounded from the outset in its applications.

*Nicolas Carnot* (d. 1832) published only one book, *"Reflections on the Motive Power of Fire"*. In it he expressed the first successful theory for the maximum efficiency of heat engines. He never lived to see his work attract much attention – people only appreciated the importance of his ideas after his lifetime. One of the first people to pick up that work was *Rudolf Clausius* (d. 1888). He stated that heat can never pass from a colder to a warmer body without some other connected change occurring at the same time. A warm body's temperature falls to match its cool surroundings. It is only possible to transfer heat from the cool surroundings towards the warm body if we input work. Whether a process like this one is spontaneous is measured by a quantity we now call entropy. Clausius' findings are a statement of what we now call the Second Law of Thermodynamics, which may alternatively be stated as saying that *"in any process, the total entropy of the universe must always increase"*.

Conversely, heat can be converted into work. This came to be accepted through the insight of *James Joule* (d. 1889) who along with *William Thomson* (d. 1907, also known as baron Kelvin*) developed an *absolute* temperature scale; one in which zero is the lowest temperature that is physically possible. Thomson was dissatisfied that the gas thermometer provided only an operational definition of temperature. Born in 1824, the same year Carnot published his estimate as −267 Celsius, Thomson used published data to calibrate his own scale. In 1848, he extended the Carnot–Clapeyron theorem to propose an absolute temperature scale, in which "*a unit of heat descending from a body A at the temperature $T°$ of this scale, to a body B at the temperature $(T-1)°$, would give out the same mechanical effect [work], whatever be the number T. Such a scale would be quite independent of the physical properties of any specific substance.*" He postulated a particular point would be reached at which no further heat could be transferred, the absolute zero temperature *Guillaume Amontons* predicted in 1702.

Thermodynamics left us four laws. Two are used often. The first law $(q - w = \mathrm{d}u)$ states that energy is conserved, so heat transfer $(q)$ and work done $(w)$ on a control mass manifests as a change in its internal energy $(u)$. The second law [17] $(\mathrm{d}s \geq \delta q/T)$ states that the infinitesimal change in entropy $(\mathrm{d}s)$ associated with a heat transfer $(\delta q)$ depends on the absolute temperature $(T)$. It has other statements, but this one suffices.

* For his work on thermodynamics, and his political positions, he was ennobled.

## *Modern oil wells and short-distance oil pipelines*

Incredibly, the Sichuan salt industry's drilling methods continued to improve until very recently. They dug the world's first kilometer-deep brine well in 1835. Although bamboo gas pipelines did not spread significantly elsewhere, Sichuan drilling methods did.

Major cities in the United States of America had coal gas or wood gas networks for lighting, but this network did not extend beyond their local vicinity. Whale oil and wax were therefore still so sought after for lighting that production peaked at 10 million gallons per year in 1854. Then, it sharply declined as whales were almost extinct. In the United States, a more common source of light was harvested by a method of carefully damaging pine trees over three years, to produce a liquid called turpentine, after which they would be cut down.

An enterprising Canadian geologist, *Abraham Gesner* (d. 1864), experimented with pitch from Trinidad and invented the process to convert bitumen into a new lighting liquid. To market his cheap, brightly burning refined liquid by coining the term *kerosene*, *"keros"* being Greek for wax, and *"-ene"* sounding familiar to consumers accustomed to buying an existing lighting liquid made from turpentine (camphene). He first commercialized his refining process to manufacture kerosene from coal oil in New York, and competitors sprung up. *

Petroleum was supplying refineries elsewhere. In 1854, oil wells supplied a refinery in Krosno, Poland built by Ignacy Łukasiewicz, and in 1856 a refinery in Ploiești, Romania processed the locally abundant oil. Gesner's refinery was successful, overcoming turpentine and alcohol interests. By 1860, the year Gesner died, dozens of competitors had oil refineries in the eastern United States, producing almost 9 million gallons of kerosene per year. The next year, the American Civil War broke out. Blockades between North and South meant whaling was disrupted, and turpentine and coal supplies couldn't be shipped.

In Ontario, the native tribes knew that gum beds at a place called Black Creek could be used to waterproof their canoes with sticky oil. *James Williams* bought a bitumen site which had failed to produce asphalt, and sought to produce lamp oil. During a local drought in 1858, he dug a well for water and struck free oil instead. Renamed Oil Springs, it was North America's first commercial oil well.

---

* A Scottish chemist filed an American patent before Gesner and forced him to pay royalties, despite Gesner inventing the process for refining kerosene first.

On the other side of Lake Erie, in western Pennsylvania, many salt wells had been dug and associated oil was known to get in the way. It was considered a *problem*; brine was pumped into a separating tank to remove the oil. Now the climate had changed, petroleum was sought after in itself. A businessman (*Edwin Drake*) struck *"rock oil"* the next year, at Oil Creek where native tribes used oil seepages for medicinal purposes. Barrels of oil were carried by horse-drawn wagon to a railroad, or loaded manually onto boats at the docks. In the 1860's, the first modern 2- and 3-inch oil pipelines were laid, moving crude from well-heads over short distances. In 1891, production peaked in Pennsylvania, and it was soon surpassed by Texas, California and other states. [42]

**Figure 47 – Barrels on boats at the port of Oil City, Pennsylvania (left) and a wooden derrick at a Pennsylvania oilfield in 1862 (right)**

Once crude oil reached a refinery, it was processed into kerosene for lighting. Other fractions of the distillation process were considered waste by-products. Gas was burned. Lighter distilled products including what we now call gasoline or petrol were simply left out to evaporate, or dumped into the nearest river.

Baku's first oil refinery was coincidentally also constructed in 1859. Footage of those early days shows a donkey rotating a horizontal wheel to extract oil, using the age-old technology people long used to raise water and brine from wells. At ground level, men stood inside the oil to shovel it. The industry developed fast. Channels were built to open tanks, then pipes to covered tanks. A kerosene plant was built in 1863 and new technologies were introduced to the refining process. By 1867 fifteen refineries were operating near Baku. The trans-Caucasian railway was built via Georgia, connecting to Batumi. Russia became entirely reliant on shipments of this oil via the Black Sea. [18]

By 1878, the region's first pipeline extended over 12 km, connecting the Balakhani field with Baku's refineries. Chemist *Dmitri Mendeleyev*, in his seminal textbook classifying the periodic table, proposed a pipeline from Baku to Batumi for export further afield. He was actively striving to catch up with Pennsylvania, seeing industries in Russia to be lagging. The enterprise was undertaken by the owners of that oilfield, the Nobel brothers. They reversed the decline, inventing oil tankers, refineries and pipelines. [19]

*Ludvig Nobel* was condemned upon his death for inventing military explosives (*"the merchant of death is dead"*). This motivated his brothers to use their oil money to fund the famous prize, leaving a better legacy. By 1900, 230 km of pipelines linked countless oil fields, transporting a million tons of oil annually, half the global supply. A new railway connected Baku with Russia. [18]

**Figure 48 – Oil barrels beside a steam train at Baku in 1884 – the railway linked it with Russia**

During the same period, in the region of Galicia which extends between what is now eastern Poland and western Ukraine near the Carpathian mountains, oil had been used as a drug for animals and as a lubricant for a long time. In 1852, *Ignacy Łukasiewicz* bought some of this oil and analyzed it with a colleague, and in 1854 moved to a town in the region to continue his work. That same year, he founded an oil well in a village near Krosno, which vies for the position of the oldest and is still in operation to this day.

He set up a kerosene street lamp, and in 1856 founded an *"oil distillery"*, which mainly produced artificial asphalt and lubricants as demand for kerosene remained low. Within three years, it burned down, and just a year later in 1860 it was rebuilt closer to Krosno.

**Figure 49 – Crude oil extraction and transportation using barrels in Galicia, Poland in 1881**

## *Flow regimes and frictional losses*

In the 1800's, naturalists in Britain, France and Germany, each motivated by their own background, were busy exploring new questions. Facilitated by these advances in materials, they revisited the science of motion and how fluids flow.

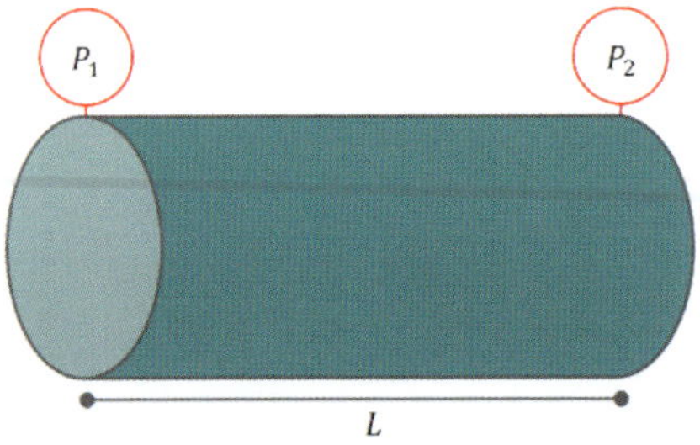


**Figure 50 – What pressure drop ($P_2 - P_1$) across a length of pipe ($L$) drives a given flow rate?**

A Berlin civil engineer, *Gotthilf Hagen* (d. 1884), undertook careful experiments in brass tubes. He discovered that at a given flow rate in the *laminar* regime, the fourth power of a pipe's radius ($R$) inversely affects the pressure drop ($\Delta P$). A Paris physicist and physiologist, *Jean Poiseuille* (d. 1869), independently discovered this same relationship, at around the same time. Hagen was more theoretically sophisticated, whereas Poiseuille's more precise measurements included liquids other than water. Crucially the latter published his findings first.

$$\Delta P \propto \frac{1}{R^4}$$

Awkward coincidences of this nature – two people making the same discovery simultaneously – are a recurring theme throughout history. Those involved were not always amicable about it. To this day, most people duly give both of these men a nod, by calling it the Hagen-Poiseuille relation for laminar flow, but the French edition of *Wikipédia* proclaims it the *Écoulement de Poiseuille*.

Euler had related the rate of change in time ($t$) of a fluid's velocity ($\vec{v}$) to its difference across space ($\nabla\vec{v}$), as well as the fluid's pressure drop ($\nabla\vec{\boldsymbol{P}}$) and the fluid's density ($\rho$). In modern notation for vector operators, his equation was:

$$\frac{\partial\vec{v}}{\partial t} + \vec{v}\cdot\nabla\vec{v} = -\frac{\nabla\vec{\boldsymbol{P}}}{\rho}$$

However, Euler's equation only holds for idealized fluids where different elements of the fluid do not exert shear forces on each other. This motivated the Mathematician *Claude-Louis Navier* (d. 1836) to draw on an analogy with elastic bodies to describe the more difficult problem: *viscous* fluids. He defined a fluid viscosity ($\mu$), using a second differential via the Laplacian operator ($\nabla^2$):

$$\frac{\partial \vec{\boldsymbol{v}}}{\partial t} + \vec{\boldsymbol{v}} \cdot \nabla \vec{\boldsymbol{v}} = -\frac{\nabla \vec{\boldsymbol{P}}}{\rho} + \frac{\mu}{\rho} \cdot \nabla^2 \vec{\boldsymbol{v}}$$

This becomes so unwieldy that it is still widely useful to solve the original Euler equation for flows where the effects of viscosity are negligible – the realm of potential flow. Unfortunately that assumption is invalid in almost all pipelines.

If these put you off, don't worry. In 1824, Navier earned a bad reputation as a civil engineer, when he failed to apply a safety margin on top of his calculations. Cracks appeared in his new bridge in Paris. It was dismantled, and he was chastised by a government committee *"for relying too much on mathematics"*.

In Cambridge, physicist and mathematician *George Stokes* (d. 1903) was a budding seventeen year old when Navier died. He saw the Hagen-Poiseuille relation was empirical, and didn't like it. Nobody knew *why* this happened, they just knew that it did. To provide a theoretical justification for how fluids flow, Stokes re-derived the equations in 1845 using continuum mechanics. He improved on Navier's work by solving two-dimensional flows. But in fast three-dimensional flows, the chaotic vortices causing turbulence proved intractable. We can only solve them by applying numerical methods, and approximations of this kind are only feasible by using modern computers.

Undeterred, Stokes went on to invent the viscometer. In a pipe, a fluid's average properties can be regarded as varying only along the length of its axis, in one dimension. A viscous flow is one in which viscosity ($\mu$) comes into play, causing it to lose a portion of its pressure ($\Delta P$) commensurate with the pipe's length ($L$) and radius ($R$), in direct proportion to the volumetric flow rate ($Q$):

$$\Delta P = \frac{8 \cdot \mu \cdot L \cdot Q}{\pi \cdot R^4}$$

We can express this relation using velocity ($v$) rather than volumetric flow rate ($Q$) by noting the cross-sectional area ($Q = v \cdot \pi \cdot R^2$). This form is most widely known today:

$$\Delta P = \frac{32 \cdot \mu \cdot L \cdot v}{D^2}$$

Putting it another way, dividing through by density ($\rho$), we can say that a new quantity which has units of height – *head loss* ($\Delta H$) – is:

$$\Delta H = 32 \cdot \frac{L}{D} \cdot \frac{\mu \cdot v}{\rho \cdot D}$$

It is equations applicable to special cases, such as this *Hagen-Poiseuille* equation, that engineers typically invoke in their daily operations, rather than the three dimensional complexity embodied in the Navier-Stokes equations. They are the lynchpin underlying the field of fluid mechanics, but hard to solve.

This simpler one-dimensional problem of flow in pipes drove progress again, this time towards quantifying frictional losses in turbulent regimes.

The Hagen-Poiseuille relation described pressure drop due to the *viscosity* of a fluid. But it didn't factor in other effects. A pipe might elevate fluid, driving it up against gravity. Conservation of energy was long established, laid out way back by Bernoulli. But nobody could yet quantify its implications for pressure drop in flowing pipes, because as a fluid flowed in practice, it seemed to *lose* a portion of its energy! Rules of thumb abounded, but progress was still lacking.

In 1845, German mathematician *Julius Weisbach* made a physically intuitive observation: that these losses scale with half-velocity-squared ($½ \cdot v^2$). If we consider the fluid to be made of tiny particles undergoing elastic collisions with the wall, the rate at which particles hit the wall is $v$ and the momentum transfer per particle also scales as $v$. It also scales linearly with a pipeline's length ($L$). In the denominator he placed acceleration due to gravity ($g$) and diameter ($D$).

$$\Delta H = f \cdot \frac{L}{D} \cdot \frac{v^2}{2 \cdot g}$$

This defines a *friction factor* ($f$) causing what is known as a *head loss* ($\Delta H$). The equation was stated in the form we use to this day. But the correlation he fitted for the friction factor ($f$) was rather different to anything we're using today:

$$f = A + \frac{B}{\sqrt{v}}$$

Though it involves coefficients ($A$ and $B$) that vary with diameter and material, it was based on a relatively small dataset; a few dozen experiments in total, many conducted by earlier French experimentalists. In fact, the equation itself was likely adapted from forgotten work by the French engineer *Antoine Chézy*.

Weisbach's comprehensive publication was a canonical engineering textbook. By 1848 it had been translated and published in America, a remarkable feat. However, his book had no apparent impact in France, despite Weisbach's links. Perhaps because his equation was based on their data, the French might have believed it provided no improvement over what they widely used already: an equation by *Gaspard de Prony*, a student of Chézy. They debated values of its coefficients, but never believed them to be a function of pipe roughness!

Interestingly, the Prony equation was not even dimensionally homogenous. In 1840 there were no electronic calculators, nor was there a modern slide rule. Prony's equation took six mathematical operations, a practical advantage over Weisbach's eight. In fact, ease of calculation remained a prime matter to consider well into the 21st century, when digital computers truly proliferated. [20] Even today, ease of use affects engineers' decisions to check a rule of thumb.

Weisbach deduced the influence of roughness, diameter and velocity on the friction factor, but was ignored. The effect of fluid properties was habitually neglected, as plain water was their only major concern. It would take almost a century of further research for friction factor to be completely defined. [20]

In 1857, *Henry Darcy* noted that *"at very-low velocity, in pipes of small diameter, the velocity increases proportionally to the gradient."* He published a new form of the Prony equation based on experiments with pipes of many diameters, over a range of velocities. Contrary to existing theory, he showed conclusively that friction factor is a function both of pipe roughness and of diameter. For this reason, friction factor is attributed to Darcy, though he never proposed it in that form. He left a tremendous legacy, perfecting the Pitot tube, and making progress in characterizing flow through porous solids such as soil.

In 1877, *John Fanning*, who founded his engineering practice following his participation in the American civil war, effectively combined Weisbach's equation with Darcy's better estimates of friction factor. Instead of attempting a new algebraic expression for friction factor, he simply published tables of values taken from French, American, English and German publications, as a function of pipe material, diameter and velocity. As he used hydraulic radius rather than diameter in his friction equation, Fanning's friction factor is defined to be a quarter of Darcy's definition. Fanning's definition is favored in fields like heat exchangers where non-circular conduits abound. [20]

Countless others made leaps in fluids during the 1800's, too many to mention in any brief summary like this. *Gustav Kirchoff*, of the famous laws in electrical circuits, also set down important equations in fluids. We will speak about just one more, then delegate to others giving the rest of that generation their due.

A fluid of low enough viscosity, flowing at a high enough velocity, through a big enough pipe is quite different in character from the flows Hagen and Poiseuille described. It doesn't follow their rules, as it is *turbulent*. In this regime, more pressure drop is required to achieve a flow rate than is predicted by the Hagen-Poiseuille relation. Attention turned to turbulence, and to many unanswered questions about the uncertain region between flow regimes which some refer to as the *laminar to turbulent transition*.

**Figure 51 – In 1883, Reynolds published his findings, accompanied with this magnificent illustration of himself with his apparatus. Although today the University of Manchester has tragically relaxed its dress code, its engineering laboratories are globally renowned.**

In Manchester, *Osborne Reynolds* investigated the nature of different flows. In 1883 his classic experiment examined the behavior of water at different flow rates, introducing a jet of dye introduced into the center of a glass pipe. In this way he could see with his own eyes the transition from a laminar regime to the onset of turbulent flow. It remains a striking demonstration to this day. The transition from the laminar flow regime to the onset of any turbulence whatsoever happens in a somewhat unpredictable fashion. An *"ill-defined, ill-behaved"* region between those two limits is known as the laminar-turbulent zone of uncertainty, and is still problematic.

Any flow can be characterized based on its ratio of inertial to viscous forces, in terms of the fluid's density ($\rho$) and its viscosity ($\mu$) as it flows at a certain velocity ($v$) through a pipe of a given inner diameter ($D$). This ratio is expressed as a dimensionless quantity known as the *Reynolds* number ($\mathrm{Re}$):*

$$\mathrm{Re} = \frac{\rho \cdot v \cdot D}{\mu}$$

* Substituting this into the Hagen-Poiseuille equation from earlier reveals that:

$$\frac{\Delta P}{\rho} = 64 \cdot \frac{L}{D} \cdot \frac{½ \cdot v^2}{\mathrm{Re}}$$

At low values of Reynolds number, viscous forces (the denominator) dominate, and we see a distinct type of behaviour which we call the laminar flow regime, so called because different layers of fluid each move in a regular manner at their own velocities. As the inertial forces (the numerator) become relatively larger, the fluid begins to experience tumbling motions, or turbulence.

## *Theory and practice*

We have jumped back and forth between theory and practice so much, we got whiplash. This history seems a bit disjointed. There is a reason for this: it was.

At the turn of the 20th century, there existed a split between theoreticians and experimentalists. Theoretical *hydrodynamics* concerned itself with inviscid flows, using velocity potential functions to represent elegant mathematical solutions of Euler's equation. They acknowledged viscous fluids violate that assumption, but nonetheless conveniently restricted their attention to exclude those effects. The big issue was turbulence. It's much easier to do theory on laminar flows, but any engineering problem involving water, air or any working fluid is almost certain to involve turbulent flows.

**Figure 52 – Workers using iron hoops to install a wooden pipeline carrying water to Seattle in the United States in 1900**

From the experimentalists perspective, this simply wouldn't do. Most practical hydraulics doesn't comply with their neat derivations, but that makes it no less real! People built pipes, pipework and pipelines, and they could see how the fluid behaved. So those engineers came up with empirical fits, meticulous charts and elaborate curves to let them get on with their designs. This begs the question: do engineers drive science, or do scientists drive engineering?

# B5 – 20TH CENTURY PIPELINES AND TECHNOLOGY

## Diesel, liquid fuels and efficiency

By the 20th century, petroleum was only widely used for lighting and lubrication. When Edison's lightbulb appeared in 1880 the oil industry wondered what next. In Munich, a student of thermodynamics, *Rudolf Diesel* was unable to graduate his class in 1879 having fallen ill with typhoid. While waiting for the next examination date, he gained practical engineering experience at a machine works in Switzerland. He graduated in 1880 with the highest academic honors and returned to Paris, where he assisted his former Munich professor, Carl von Linde, with the design and construction of a modern refrigeration and ice plant. Becoming director of the plant one year later, he married, and continued to work for Linde, still the world's largest industrial gas company.

Diesel did not like that the theoretical and practical constraints on a steam engine's efficiency wasted 90% of the energy available in a fuel. He moved to Berlin, striving to build something more efficient. He built an ammonia engine, which exploded and almost killed him. Then, he tested the strength of iron and steel cylinder heads, one of which exploded. He was hospitalized, and suffered from chronic health problems that affected his eyesight. Still he pressed on. In 1893, he published a treatise, confidently entitled *"Theory and Construction of a Rational Heat-engine to Replace the Steam Engine and The Combustion Engines Known Today"*. By the end of the summer, he realized his theory was wrong, filing another patent application for his corrected theory that year. The director of MAN SE, in Augsburg, gave Rudolf Diesel the opportunity to test and develop his ideas. The first successful Diesel engine ran in 1897. He obtained patents for his design in many countries, including the United States. *

By 1913, the world was on the brink of its first devastating Great War. Diesel went to Antwerp and boarded a steamer to London. He took dinner and retired to his cabin at 10 p.m. In the morning his cabin was empty, and his bed hadn't been slept in. His hat and neatly folded overcoat were discovered beneath the afterdeck railing. His biographers argue he committed suicide. Another theory suggested he was murdered, because he'd refused to grant Germany exclusive rights to his invention, then boarded a steamer to meet representatives of the British Royal Navy. Evidence is limited, and his death still remains unresolved. The First World War was the first war fueled by oil. Petroleum became the lifeblood of military force, a strategic asset for entire economies.

* He designed his engine to run on vegetable oil. Many diesel engines can still do so.

## Mach and the speed of sound

Austrian physicist and philosopher Ernst Mach (d. 1916) spent his early career conducting experiments into the nature of light. It is from that background that he embarked to explore fluid mechanics. Specifically, he was interested in supersonic fluid mechanics: objects moving faster than the speed of sound. His team deduced and experimentally confirmed shock waves that occur as a projectile crosses the sound barrier. The impact of their investigations would prove wide reaching. His son, *Ludwig Mach* (d. 1951), continued his legacy.

The speed of sound varies in different ambient media. In simpler terms: each fluid has a different sound speed, and that sound speed depends on the local conditions of the fluid, its pressure and temperature. The Mach number ($\mathrm{Ma}$) is defined as the ratio of a fluid's velocity ($v$) to the local speed of sound ($c$):

$$\mathrm{Ma} = \frac{v}{c}$$

This has lots of applications, including a couple in pipelining. Although all gases are compressible, sometimes they are imperceptibly so, so we can treat them as incompressible in certain conditions.

## Joukowsky and water hammer

It is at this period in history that powered flight went from an outlandish idea people scoffed at to a reality of urgent strategic importance. Influenced by the Machs (father and son), *Nikolay Zhukovsky* founded Russian aviation, establishing the world's first aerodynamic institute in Moscow in 1904. As pipeliners, we have him to thank for an important hydrodynamic equation describing a phenomenon called *water hammer* or *hydraulic surge*. It is known as the *Joukowsky equation.* * A change in fluid velocity ($\Delta v$) causes a change in pressure head ($\Delta H$) dependent on speed of sound in the fluid ($c$):

$$\Delta H = -\frac{c \cdot \Delta v}{g}$$

The magnitude of the pressure surge ($\Delta P$) can then be calculated based on the fluid's compressibility and density. And although some pipeliners are not familiar with the Joukowsky equation, this result is helpful for validating transients and surge analysis.

* Transcribing from Cyrillic can be inconsistent.

# The value of the friction factor

In 1904, German aerospace physicist *Ludwig Prandtl* presented a ground-breaking paper: *"On the Motion of Fluids in Very Little Friction"*. He described how a *viscous boundary layer* causes flow to separate. This discovery finally reconciled mathematical *potential flow* with experimental *viscous flow*. It was so influential that he was appointed director of the technical institute at the University of Göttingen. He led it through unprecedented advancements in aerodynamics, leading the world for decades until the end of World War II. His focus was on aerodynamics, examining the compressibility $(\kappa)$ of a gas as it approaches the transition from subsonic to supersonic speeds (Mach 1).

In pipelining, we need to determine whether convection or conduction is the limiting heat transfer mechanism outside a pipe immersed in fluid. This is a somewhat niche application of a very important concept. The Prandtl number $(\mathrm{Pr})$ is a dimensionless ratio of momentum diffusivity to thermal diffusivity, both in SI units of $\mathrm{m^2/s}$. Given the specific heat capacity $(c_P)$, dynamic viscosity $(\mu)$, and thermal conductivity $(\lambda)$ of a fluid, the density $(\rho)$ cancels as follows:

$$\mathrm{Pr} = \frac{\text{momentum diffusivity}}{\text{thermal diffusivity}} = \frac{\mu/\rho}{\lambda/(c_P \cdot \rho)} = \frac{c_P \cdot \mu}{\lambda}$$

Whereas a Reynolds number $(\mathrm{Re})$ is associated with a scale variable such as pipe diameter, a Prandtl number $(\mathrm{Pr})$ involves no such length scale in its definition. It only depends on the fluid and the state of the fluid, so it is tabulated as a property of a fluid. He applied *similarity theory* to experimentally establish that in smooth pipes around $4\,000 < \mathrm{Re} < 80\,000$, friction factor $(f)$ depends on Reynolds number:

$$f = \frac{0.3164}{\mathrm{Re}^{0.25}}$$

This is now called the Blasius formula. *Paul Blasius,* one of Prandtl's students, laid down a mathematical basis for friction in the laminar boundary layer, showing that resistance to flow through smooth pipes is a function of Reynolds number both in the laminar regime and in the smooth-pipe turbulent regime.

Prandtl appointed a Georgian student, an outstanding *Johann Nikuradse*, as head of his department. Together with his Hungarian colleague, *Theodore von Kármán,* they furthered our understanding of friction in the turbulent regimes. Nikuradse undertook many experiments. The most famous carefully measured friction experienced by a turbulent fluid as it flows through a rough pipe. Using grains of sand to adjust the roughness of the pipes' inner wall, he discovered that rougher surfaces give rise to more friction, imposing a greater head loss.

Figure 53 – Does roughness protrude through the viscous boundary layer in a turbulent flow? Answering that question leads us to distinguish between rough- and smooth-pipe turbulence.

He confirmed that in the range of moderate Reynolds numbers, resistance factor stays the same, regardless of whether a pipe is rough or smooth. The reason is that rough projections from the pipe wall remain entirely contained within the laminar boundary layer. In plainer terms: the bulk mainstream flow doesn't *"see"* the roughness of the wall. This regime is *smooth-pipe turbulence.*

Then, as he looked at more substantial Reynolds numbers, he observed a distinct *transition between smooth- and rough-pipe turbulence*, a range within which resistance increases as Reynolds number ($\mathrm{Re}$) is increased. The reason given is that the thickness of the laminar layer is in this range has the same order of magnitude as the projections. In plainer terms: the nature of a flow affects how much wall roughness it interacts with.

Finally, in *rough-pipe turbulence*, resistance factor is independent of Reynolds number, because *all* projections extend through the laminar boundary layer. Resistance factor in this regime depends only on the relative roughness:

$$\frac{1}{\sqrt{f}} = 1.14 - 2 \cdot \log\left[\frac{\varepsilon}{D}\right]$$

The government hounded Nikuradse for soviet espionage. He was dismissed from his post as head of department, despite Prandtl's protests and his own connections within the Nazi party. Von Kármán emigrated, and was appointed director of the Guggenheim aeronautical laboratory in California. Blasius, as far as we could tell, was spared any such trouble, and continued teaching in Hamburg until he died in 1970.

The transition from laminar flow to rough-pipe turbulence as explored by Nikuradse in 1933 was based on uniform roughness. His sand coated the entire interior surface of the pipe. His data could be explained by the interaction of the pipe's roughness with the fluid's boundary layer. In 1937, two civil engineers in London, *Cyril Colebrook* and *Cedric White*, showed that flow through pipes with *non*-uniform roughness did not make its transitions from laminar flow to rough-pipe turbulence that way. In 1939, they coined a formula to correlate the friction factor of commercial pipes within the transition region:

$$\frac{1}{\sqrt{f}} = 1.14 - 2 \cdot \log\left[\frac{\varepsilon}{D} + \frac{9.35}{\mathrm{Re} \cdot \sqrt{f}}\right]$$

This formula was to attain terrific popularity, but remains a little controversial as the reason for this was its *adequacy*, not that it ever claimed to be excellent. In 1942, American hydraulician *Hunter Rouse* was excited but exasperated. *"These equations are obviously too complex to be of practical use. On the other hand, if the function which they embody is even approximately valid for commercial surfaces in general, such extremely important information could be made readily available in diagrams or tables."* He attempted a diagram, and sought to better define the boundary from the transitional regime to turbulence.

As Rouse presented all this to his audience, *Lewis Moody* listened attentively. The Rouse diagram lets us look up flow rate at a given head loss and diameter. But we know the flow rate! It's cumbersome for the common task of looking up head loss at a given flow rate and diameter. So Moody took it upon himself to redraw it, *"his aim merely being to embody the now accepted conclusion in convenient form for engineering use."* Despite acknowledging the previous researchers, and reproducing figures from Colebrook and Rouse, the famous diagram was attributed to him. Rouse hated this. He writes – referring to himself in the third person:

> *"After the Conference, Lewis Moody of Princeton suggested using the latter variables ($f$ and $Re$) as primary rather than supplementary, as in the past, but Rouse resisted the temptation because he felt that to do would be a step backward. So Moody himself published such a plot, and it is known around the world as the Moody diagram!"*

Apparently, this is one of few instances in which Rouse opted for an exclamation mark.

The Moody diagram has not been modified since 1944. Fifty years later, in 1994, *Frank White* estimated that the Moody chart is only accurate to within ± 15%. Yet today, its results are widely applied as if they were laws of nature, unquestioned by many engineers, despite that simple look-up being inappropriate in some conditions.

This raises questions about the implications of this uncertainty. A sensitivity analysis might quantify the repercussions of a pivotal equation being so far off.

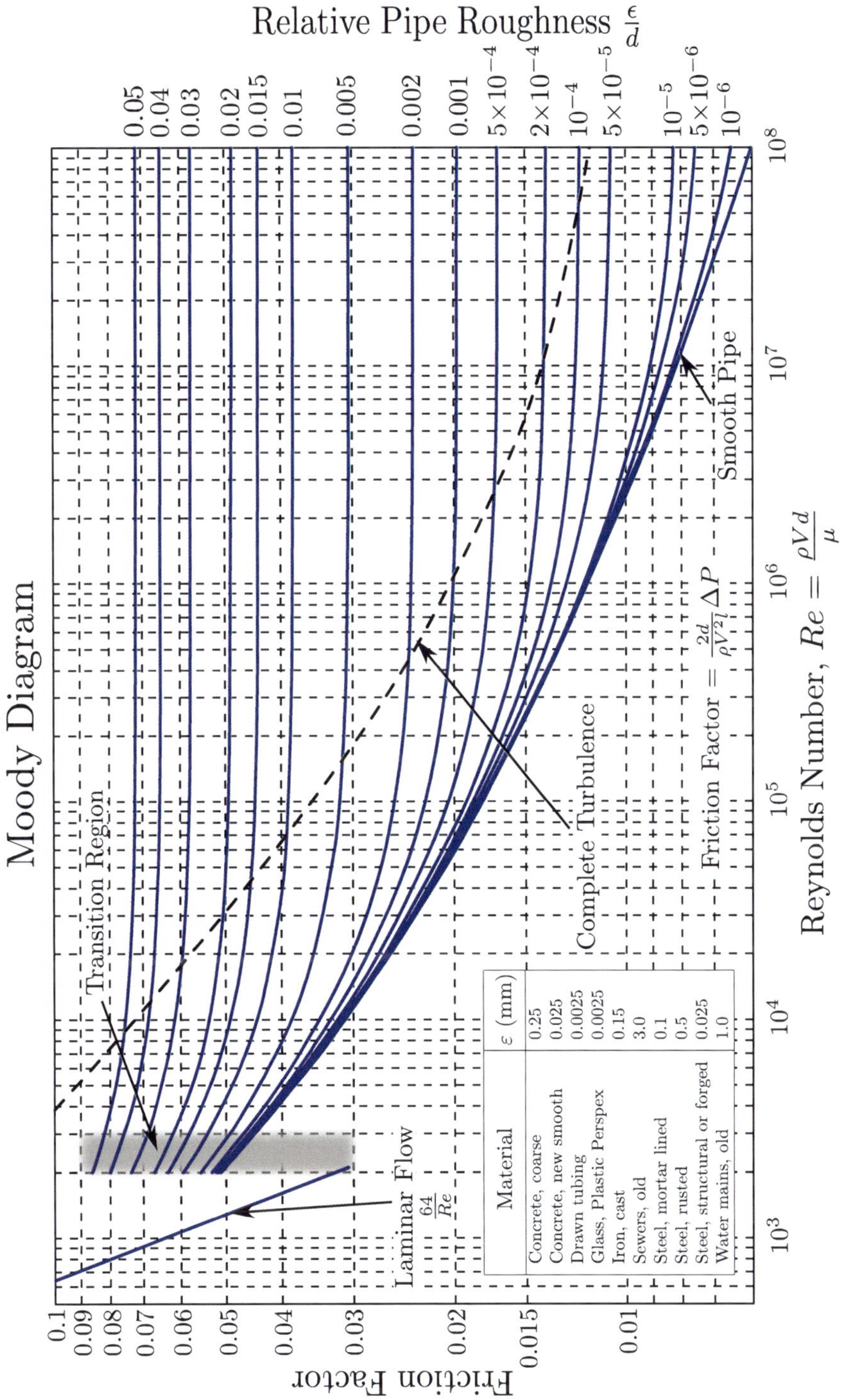


**Figure 54 – The Moody Diagram remains in widespread use essentially unchanged since 1944**

**S Beck and R Collins at the University of Sheffield / Wikimedia Commons / CC BY-SA 3.0**

# Scale models

Given the controversy around the accuracy of the Moody diagram, one might ask: could we not simply build better experiments than we did in the 1940's?

To make a scaled-down version of a pipeline in our laboratory, we design it to behave like the real system. To this end, we strive to design the scale model so that it shares the *dimensionless quantities* characterizing the real pipeline.

For frictional forces to be the same, *Reynold's* number ($\mathrm{Re}$) must be kept the same:

$$\mathrm{Re} = \frac{\rho \cdot v \cdot D}{\mu}$$

For pressure forces to work the same, *Euler's* number ($\mathrm{Eu}$) must be kept the same:

$$\mathrm{Eu} = \frac{\mathrm{d}P}{\rho \cdot v^2}$$

So to shrink our length scale – that is, the pipeline diameter ($D$) – we need a fluid with higher density ($\rho$) at more intense pressure ($P$). By this principle, experimentalists could generalize flow in pipes, and it is the same principle a wind tunnel uses to model aircraft aerodynamics. Unfortunately, it suffers the same limitations: this scaling approach fails to properly capture *thermal* effects.

When pipeline simulators became available, engineers could avoid the inconvenience of constructing scale models, and were finally able to conduct virtual experiments capable of painting a more accurate picture of what truly happens in a pipeline.

By the mid-20$^{th}$ century, there was no shortage of operational data from real pipelines to validate the predictions of friction factor correlations.

## Long-distance oil pipelines

In 1907, after a decade of construction, Mendeleev's long-distance pipeline was finally complete. 8-inches in diameter, it transported nearly a million tons of kerosene per year from Baku to Batumi, on the coast of the Black Sea, for onward export by oil tanker. Extending almost 900 km, equipped with 16 pumping stations, and rising 800 meters above sea level, it was the longest in the world. [18] The pumps along the pipeline were *steam-powered*, supplied by an American company, Worthington of Brooklyn. [19]

The Soviet Union invaded in 1920, securing oil to fuel Lenin's industrialization. In 1925, the pipeline's service was changed from kerosene to crude oil. A parallel 10-inch pipeline was soon built, transporting more crude oil to a refinery at Batumi. Newer, German-made pumps driven by diesel engines were fitted, requiring fewer pumping stations. Baku continued to supply Russia with oil for decades. In the 1940's, for fear of falling into the hands of the enemy trying to invade the area during the epic battle for Stalingrad, this large pipeline system was *dismantled*. It was hastily re-assembled further east during that war, between the Volga and the Urals.

By the 1900's, American pipelines had also grown from early 2 or 3 inch lines to a standard pipe size of 8 inches. Any larger diameter was found potentially liable to split at the seams. But an eight-inch pipeline could only deliver 20 000 barrels per day. When Texas oil boomed in the 1920's this became a bottleneck: six times less than the crude oil consumed by a major refinery in 1930. Initially a work-around was found by *looping*. A loop is a second adjacent pipeline laid alongside the first. In the 1930's, advances in technology let manufacturers draw *seamless* steel pipelines at diameters of 12 inches or more. They called these *"big-inch"* pipes. Pipelines could finally be built to carry crude oil and refined products over long distances. Then, the great depression struck. Over a decade only 10 000 miles of pipeline were laid in the US. [42]

When the US entered the Second World War in the 1940's, Nazi U-boats attacked dozens of oil tankers off the Atlantic coast. These attacks were visible from the coastline and witnessed by many horrified onlookers. They disrupted crucial supplies to the north-eastern seaboard and its military industries. Something had to be done, so the government decided to build a pipeline from Texas to New York. Eleven private oil companies formed the *War Emergency Pipelines* in 1942. Construction started the next day. Within a year, a fifty-mile slug of water was sent through for hydrostatic testing to check for leaks. [42]

At the time, the *Big Inch* pipeline was the largest, longest pipeline ever built. 24-inches in diameter and 1 200 miles long, with dozens of pumping stations spaced every

50 miles moving 335 000 barrels of crude per day, travelling 40 miles per day over many weeks. A parallel pipeline, *Little Big Inch,* was built to share the original pump stations. Rubber balls separated batches of refined products: gasoline, kerosene, diesel fuel and heating oil. Big Inch and Little Big Inch were arguably the first true long-distance pipelines. Laid across swamps, forests and the Allegheny mountains, they traversed hundreds of rivers, creeks and lakes, crossing the Mississippi in a trench blasted across its riverbed. [42]

These are the decades in which modern innovations to control sophisticated operations became widespread, where human interaction might be impossible or dangerous. During the industrial revolution, water *"hydraulically"* actuated valves and equipment. By the 1920s air was used *"pneumatically"*. As process plants required higher pressures, actuation used more sophisticated electric, direct-gas or gas-over-oil designs. These improved until in the 1950s, electro-hydraulic actuators were in use for critical fail-safe applications, and high-pressure gas actuators were used for the high pressure pipelining industry. 21

Pipelines appeared everywhere. Up until the 1960's, high-voltage power cables were buried in high-pressure fluid filled (HPFF) pipes. To prevent sparks, cable was routed through an oil-filled pipeline. Pumping stations carefully circulated oil back and forth between two tanks to keep it cool. And to perform maintenance, a section was drained not with valves (in light of the cable), but by sending liquid nitrogen down a pair of access shafts either side, to freeze the oil solid.

## Vapor pressure and slack

In routine operation, the oil around these power cables was maintained at a pressure of 14 atmospheres or so. This is because power cables get hot, and so the oil did too.

Temperature greatly affects the volatility of any liquid. That is to say: the warmer an oil is, the more of it will be in a vapor phase, and the more likely it is that vapor pockets might form. For this reason, pipeliners either maintained enough pressure so that their liquid never fell below its vapor pressure, especially on a warm day at a high point, or else expect their liquid to go partially or entirely into the vapor phase.

This transition of liquid to its vapor phase was characterized on a theoretical basis by Clausius and Clapeyron, both of whom had worked in the mid-1800s as we saw earlier. Their equation was soon improved for use in practical fluid applications by an obscure naval engineer, Louis Charles Antoine (d. 1898), fitting parameters ($A$, $B$, $C$) to relate vapor pressure ($P$) to temperature ($T$) via:

$$\log[P_{\text{vap}}] = A - \frac{B}{T + C}$$

Although this had already been well-known for decades when long-distance oil pipelining emerged, it was easier said than done. A tall mountain peak sits above a tall column of liquid, meaning that the pressure at the base must be exceedingly high if every high point in an oil pipeline is to be kept above its vapor pressure. Pumps could not necessarily achieve such pressures in the early days, or if they could the strength of the pipeline and its joints could not withstand it. Even granting that, the cost of all that pumping was sometimes prohibitive. This all meant that vapor pockets formed along certain sections of a hilly oil pipeline, especially during shut-in conditions; a phenomenon called *column separation*. Operators learnt how to carefully start up their pipeline to cope with that reality, and even how to operate in what we call *slack-line flow* wherever *tight-line flow* was deemed to be inconvenient or impossible to achieve.

# Van der Waals, real gases and changes of phase

Just as many people wrestled to grasp the specifics of how a liquid changes into a vapor, others were trying to understand how gases behave under high pressure as their density changed significantly differently to the *"ideal"* gas law, and how below a certain temperature and / or above a certain pressure the least volatile components of a gas mixture condense into a liquid phase.

In 1910, *Van der Waals* received the Nobel prize for physics. He addressed a shortcoming of the ideal gas law: it under-estimates density at high pressures. To do so, we introduce a *compressibility factor* or a *supercompressibility* $(Z)$*:*

$$\rho = \frac{m_r \cdot P}{Z \cdot R \cdot T}$$

This quantity is *not* a constant. Its value varies with pressure, temperature and composition. By adjusting for the volume of molecules and the attractive force between them, he wrote an equation of state to capture this effect:

$$\left(P + a \cdot \left(\frac{\rho}{m_r}\right)^2\right) \cdot \left(\frac{m_r}{\rho} - b\right) = R \cdot T$$

Which is equivalent to saying that the supercompressibility varies as follows:

$$Z^3 - (1 + B) \cdot Z^2 + A \cdot Z - A \cdot B = 0$$

Where $A$ and $B$ can be determined without too much trouble.

He also observed that all fluids have about the same compressibility factor (and other thermodynamic properties) if we divide their pressure by the critical pressure, and divide their temperature by the critical temperature. These are referred to as reduced pressure and reduced temperature, and this principle is called *the law of corresponding states*.

## Short-distance natural gas pipelines

During the first world war, advances were made in welding and widely used in shipbuilding instead of older riveting techniques. During the great depression that followed, those welders found little demand for shipbuilding, but their skills were put to good use for a new application: gas pipelines. In the 1920's, the Texas oil boom produced natural gas as a pesky by-product. In 1925, a 200 mile bolted natural gas pipeline was rebuilt in Texas using acetylene welded pipe. Within a few years, electric arc welding developed, a method that eliminated overlapping of pipes, and that halved the workload. Alloy steels, ditching machines and gas compressors all improved too, and by 1931 the first thousand-mile gas pipeline was being laid from Texas to Chicago. [42]

In 1918, the American National Standards Institute (ANSI) was founded. The Gas Processors Association (GPA) began in 1921. At first, the industry was disorganized. There were over 100 different specifications for gasoline and no meaningful tests for determining product quality. A few plant operators formed an association, and adopted the first standard specifications. They developed simple, reproducible test methods for determining relative density, vapor pressure, and other tests to define their industry's products. Nearly all natural gasoline sold since 1922 and all liquefied petroleum gas (LPG) sold since 1930 have been traded on the basis of the GPA's specifications.

Natural gas burns cleaner and offers twice the calorific value, while costing only a third of the price of the old towns gas manufactured laboriously from coal. But its rollout was hindered by the cost of traversing the sheer distances that separated wellheads from urban consumers. There was also uncertainty in the early days around how substantial a gas reserve actually was. Nonetheless between 1927 and 1936, at least twenty-nine major American cities converted their towns gas distribution grids to natural gas, and an extensive distribution network *"spidered"* its way up much of the country. In Chicago, vested interests kept coal gas alive by blending it with natural gas. But still in 1935, more natural gas was being wasted (4 375 billion cubic feet – BCF), vented or flared off to thin air, than was being consumed for any useful purpose (3 520 BCF). [42]

As gas was now being shipped across many jurisdictions, local municipalities found themselves unable to curb all manner of rampant anticompetitive and monopolistic practices by gas transmission companies. It was the president himself who intervened. Franklin Roosevelt's New Deal strove to overcome the economic depression, and for the first time, the US government started regulating the natural gas industry to that end in 1938. This allowed pipelines to be built, rapidly expanding the natural gas delivery network. A federal agency took responsibility for regulating rates of interstate natural gas transmission pipelines. Gas production and gathering pipelines were exempt from this regulation, so the companies extracting gas argued at the supreme court that natural gas sold at the wellhead was therefore exempt. Consumer groups campaigned otherwise. The ruling was for price controls that lasted 40 years.

How much does it cost to pipeline natural gas?

In 1937, the Hindenburg set ablaze as it moored to land after crossing the Atlantic. The Zeppelin was cloth bags full of hydrogen at atmospheric pressure. We had many of our modern gases for decades, but could not pressurize them. In fact, we could not even properly contain on-board liquid *diesel* at pressure. Humanity had advanced in our science of motion, but lagged in our materials.

The enabling technologies for pipelining gases of any sort over long-distances happened in the realms of welding, materials, and compressors. These developments took place decades later than liquid pipelines. Although low pressure *towns* gas distribution networks existed in urban areas since the 1800's, *natural* gas could not reach customers across long distances until we had high-pressure *transmission* pipelines. Those would finally appear in the late 1900's.

## Long-distance natural gas pipelines

When the war ended in 1945, Big Inch and Little Inch became war surplus, and *"an intense debate followed".* Although natural gas pipelines now dotted many parts of that country, none yet extended far enough to reach New York. Homes and industries throughout the north-eastern states still relied on locally distributed towns gas, manufactured from coal at great expense. Fifteen million people relied on it for cooking and hot water, though few were using it to heat their homes. Housing would be built for veterans coming home, and advances in compressors now facilitated long-distance gas transmission. The time and place were right for a talented assistant to the president of Standard Oil to propose converting the idle lines for cheap, plentiful natural gas, which consumers preferred over towns gas. This became the norm in major cities. [42]

The change met with *"vociferous objection"* from the United Mine Workers, and of course a *"firestorm of resistance"* from coal gas manufacturers. Coal miners were led to strike by their tenacious union leader, who so infuriated president Harry Truman that the President made a personal note for the record of his version of events – one that was thoroughly unflattering. Within a couple of years, the miners strikes abated, and the lines still deliver natural gas today. [42]

Converting liquid pipelines to natural gas service is unusual, but not unheard of; the same was done to a fair number of pipelines around the world. In Britain, the driving factors were a little different. In the winter of 1952, a *"killer fog"* caused by pollution killed thousands of Londoners. They studied the cause of this tragedy, and concluded that they had to find a better way of heating their homes than burning coal. First they expanded the coal gas network to provide more homes with gas for heating – specifically, this *"towns gas"* was a manufactured mixture of gases consisting partly of carbon monoxide (CO) and partly of hydrogen ($H_2$).

Within a decade or two, natural gas fields were found in the North Sea. In the 1960's and 1970's, Britain and Europe switched from coal gas to natural gas. The absence of a high-pressure transport grid back then made such transition far less complicated than it would be today. The grid was divided into different geographic sectors; so it was that *"literally overnight"* the replacement of gas type was carried out in a particular sector. In the subsequent week, all end-use applications were modified or replaced. Natural gas became the energy carrier in Britain and much of western Europe.

In 1992, the American Gas Association (AGA) issued their "*Compressibility Factors for Natural Gas and Other Related Hydrocarbon Gases"*. Implemented in today's pipeline simulators, it is an equation of state referred to as AGA-8.

# Human computers, digital computers and risk

When Hunter Rouse (who we met earlier, and didn't care much for Moody) wrote his *History of Hydraulics* in the 1950's, he did not so much as mention computers. [22]

> *"Prior to the 1960's, most pipeline simulation was done by hand calculation using slide rules and log tables… or lookup tables. Most of this work used empirical correlations… there was no friction factor to calculate [nor] iterations to find the answers."* [23]

This would all change during the cold war. The way computers came to design, monitor and operate pipeline infrastructure is fascinating, but is not one we can delve into here. Suffice to say, by the 1980's, digital computers were the way this critical infrastructure came to be operated. This raised questions about the vulnerability of a pipeline system, questions the world still grapples with today. In an age where a pipeline's operation relies on software that must gather and continuously transfer unprecedented amounts of sensitive data, experts must ensure their digital infrastructure is safe while more connected to the outside world than ever. This comes into sharp focus when it is being used to operate physical infrastructure.

# B6 – THE INFORMATION AGE

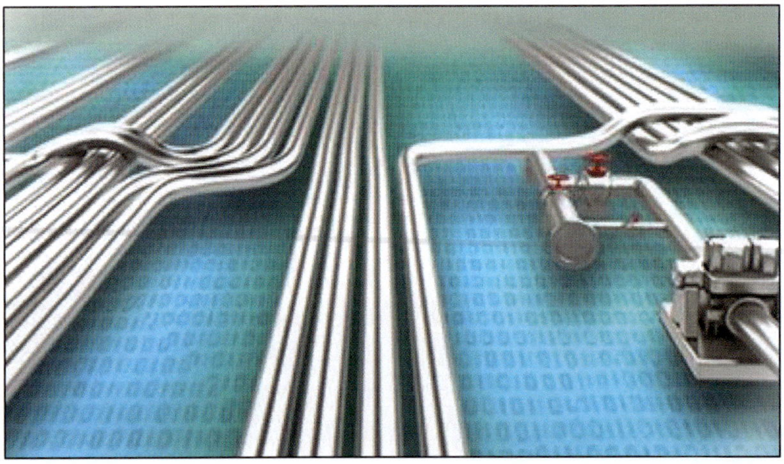

Figure 55 – Modern computing has heralded a new age that digitally transformed pipelining

Capable software engineers worked around the limitations of early computing by carefully designing software to be lightweight. In the following decades, computers advanced rapidly, at a fast accelerating pace from 2000 until today.

Many pipeline simulation software vendors struggled to keep up with these advances. The pace of change in computing power, programming paradigms, and the wider tech industry is at the opposite end of the spectrum compared to conservative, slow-to-change industries like water or oil and gas. This poses special challenges to companies in that not only is change needed constantly for inevitable patches, fixes and features, but the entire programming language of some older pipeline simulators quickly became outdated during those years.

A few companies started offering several simulators – but maintaining older software because a pipeline operator sees no reason to upgrade meant less time was spent on newer software. That ongoing maintenance became more expensive over time as no developers could be found speaking in the forgotten tongues of now obscure 1980's programming languages. And it became more *pointless* too, as what they were told to maintain became laughably antiquated.

There is a famous *"S-curve"* people talk about; a principle that rationalizes whether incremental improvements are worthwhile, or if a technology has reached maturity and a step-change is about to overturn the status quo in a new technology revolution.

For decades, existing pipeline simulators became faster year-on-year without any expenditure by their makers, as computers themselves became faster. In the 2020s processor speed has stalled, but processors have become cheap and plentiful. At the start of the century, it was not unusual for larger pipeline enterprises to have a specialized simulation department. Two decades on, it's now not unusual for a *new graduate* to have dabbled in simulation. Pipeline simulation software packages are accessible to more users than ever.

Live on site, advanced *state estimators* accurately predict mid-line conditions viewed as profiles, reporting points and virtual meters. Instrumentation, smart and dumb, is monitored better than ever, its data crunched, analyzed and thoroughly validated. When a meter goes bad, it is rapidly identified, while the wider model remains stable and carries on performing. Specialized models for batch tracking applications have also come leaps and bounds in recent years.

Ongoing development work is focusing on the future of pipeline simulators by re-architecturing the software to take full advantage of cloud computing. Look-aheads are automatically forecasting more scenarios, more accurately, and longer into the future. They are today finally able to assist the operator with suggested ways of averting an anticipated disruption or violation.

Advanced look-aheads work a lot like optimizers, making recommendations about the best course of action amongst countless possible strategies. In fact, optimizers themselves are being deployed alongside look-aheads, harnessing computing power to make the operation of pipelines smarter than ever before.

Pipeline simulators are offering the teams and enterprises operating pipelines a greater insight into their operations than could ever have been imagined. What we're doing is at the cutting edge, pushing the bounds of what is possible. Despite this, if we really understand it, we should be able to *"explain it to a five year old"*. In the next chapter, we attempt to do just that. We assume the five year old is bright.

# Appendix C – A primer on pipelines

*"We choose to do [these] things, not because they are easy, but because they are hard"* – John F. Kennedy

## C1 – Understanding our domain

What warrants our attention? A well-defined problem statement helps us stay focused. It is otherwise tempting to end up going every which way, getting side-tracked with accounting tools, logistics and a host of clever data processing. The myriad of use cases for pipeline simulation can become quite dizzying, but by and large, this is how we go about thinking of what matters: What is it about a pipe and a fluid that is actually important to us?

To build sound software, and good models to start simulating, we should familiarize ourselves with our *domain*. So we start with a dedicated chapter about pipelining. A basic primer on how pipelines work, and how pipeliners design and operate them. A lot of this isn't fun, being less familiar to people than astrophysics or rocket science. We dig into the details of fluid calculations in later chapters; first we equip ourselves with a basic roadmap before adding more to our toolkit. The tedium involved in a lot of this is precisely what makes number-crunching simulators a compelling value proposition.

A pipeline is one of the simplest applications of fluid mechanics in modern engineering, being essentially one-dimensional. In spite of this simplicity, state-of-the-art computer simulations control and predict its behavior. Pipeline simulator systems are no less sophisticated than those widely deployed in aviation, though sadly not spectacular enough to watch on the glamorous big screen. Geniuses across the ages worked on the physics of fluids, including Noble prize winners, and one credited with casually and inadvertently founding an entire separate field: the kinetic theory of gases. What is so hard about pouring fluid into one end, so it comes out of the other? Let's find out.

# C2 – STRESS IN PIPES AND PRESSURE VESSELS

*"Although some accidents were due to criminal practices such as tying down safety valves, most were caused by lack of proper calculations. This is a pity because the basic calculations to determine the stresses in simple pressure vessels are [so easy] indeed that nobody has ever bothered to claim the credit for originating them".* [14]

The skin or shell of a pressure vessel performs two functions. It contains the fluid, and it carries stresses set up by any *internal pressure* ($P$) we apply to it.

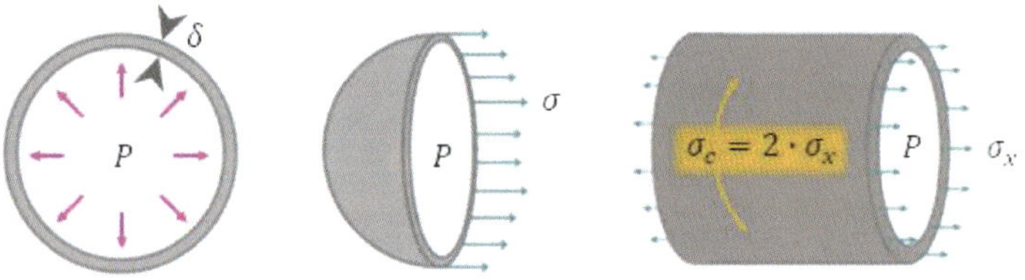


Figure 56 – Tensile stress in a spherical pressure vessel, and tensile stresses in a pipe

In any pressurized sphere or cylinder, the stress acting outwards in the radial direction is negligible; we're concerned with the stress of *tension* ($\sigma$) acting in the other two spherical co-ordinates. Slicing a spherical pressure vessel of a radius ($R$) in half we see the tensile stress acting along its *wall thickness* ($\delta$) is:

$$\sigma = \frac{R \cdot P}{2 \cdot \delta}$$

Taking a slice of a cylindrical vessel (a pipe) in the longitudinal direction ($x$) we see the same stress as in a sphere acting along its walls – an *axial stress* ($\sigma_x$):

$$\sigma_x = \frac{R \cdot P}{2 \cdot \delta}$$

Slicing our cylindrical vessel again in the circumferential direction ($c$), we see that the *hoop stress* ($\sigma_c$) acting in that direction is *double* the axial stress ($\sigma_x$):

$$\sigma_c = \frac{R \cdot P}{\delta} = 2 \cdot \sigma_x$$

We care about the hoop stress because that is the component of stress that leads to pipe failures. An operating pipeline must maintain pressures keeping hoop stress below the yield stress of the steel by a large, usually government-mandated value. In addition, as we will see in the section on wave propagation speeds, pipe expansion interacts with pressure surges in the fluid in a manner that can significantly slow down transients.

# C3 – How pipes are like electric circuits

## Flow in the circuit analogy

People like to draw on an analogy between direct-current (DC) electrical circuits and the hydraulics of an incompressible liquid like water flowing through a pipeline. It also extends to heat transfer, mechanical devices and other systems. [24]

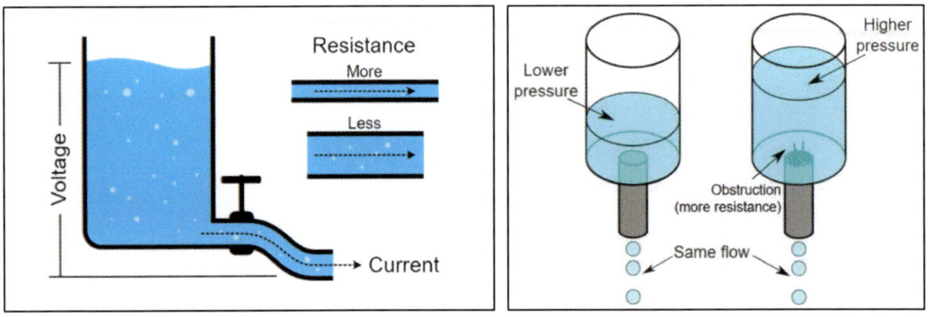


**Figure 57 – A difference in *head* causes water to flow across a *resistance*, like electric current**

Pipeline simulators bring together people from all kinds of backgrounds. Some of us are initially more familiar with electric circuits than fluid mechanics. So let's explore how far we can take this analogy, and where exactly it falls short. A wide *pipe* full of an incompressible liquid, like water, transmits pressure immediately. A *block valve* opens or closes to allow or stop flow like an on/off switch does for electric current. A *check valve* stays closed unless a forward threshold of pressure drop occurs, preventing reverse flow much like a diode.

## Potential and resistance in the circuit analogy

Pumps raise pressure, as batteries raise voltage (engineers call it *potential*). Potential difference is analogous to *pressure drop*: it is an *"across-variable"*, driving a *"through-variable"* (current, or flow) through a resistance. A resistor is analogous to the resistance to flow presented by a section of pipe or a valve or fitting. That might be a constriction – a smaller diameter – or a long stretch. Abstracting a pipe's resistance is a basis for leak location algorithms.

| Domain → Concept ↓ | Direct current | Heat conduction | Liquid hydraulics | Mechanics (I) |
|---|---|---|---|---|
| **Potential (*"across"*)** | Voltage | Temperature | Pressure | Velocity |
| **Conserved quantity** | Charge | Heat | Volume * | Momentum |
| **Flow *("through")*** | Current | Rate of heat transfer | Volumetric flow rate | Force |
| **Flow per unit area** | Current density | Heat flux | Velocity | Stress |
| **Impedance** | Resistor | Thermal resistivity | Flow coefficient | Inverse of friction |
| **Storage** | Capacitor | Heat capacity | Tank | Mass |
| **Inertia** | Inductor | - | Flywheel or long pipe | Compliance † |
| **Energy** | *"Watt-hours"* | Heat (*"Joules"*) | Flow work (*"Joules"*) | Force times distance |
| **Power** | *"Wattage"* | Rate of heat transfer | *"Horsepower"* | Force times velocity |
| **On / Off** | Switch | - | Block valve | - |
| **Non-return** | Diode | - | Check valve | - |
| **Motive force** | Battery | - | Pump | - |

**Table 1 – Analogous concepts in direct-current electric circuits, thermal conduction, liquid hydraulics and mechanical systems used for energy storage. Blank entries are rarely used.**

* If the density of a fluid varies, like in a gas, we may only use *standard volume* or *mass*. If the composition also varies, we may only use *mass*.

† The inverse of a spring constant. The reader should be aware another mechanical analogy swaps velocity and force around.

# Capacitors in the circuit analogy

A capacitor stores energy electrically, much as a battery stores it chemically, pumped water stores it mechanically, and molten salt stores it thermally. Large supercapacitors are used alongside all these other methods for grid balancing.

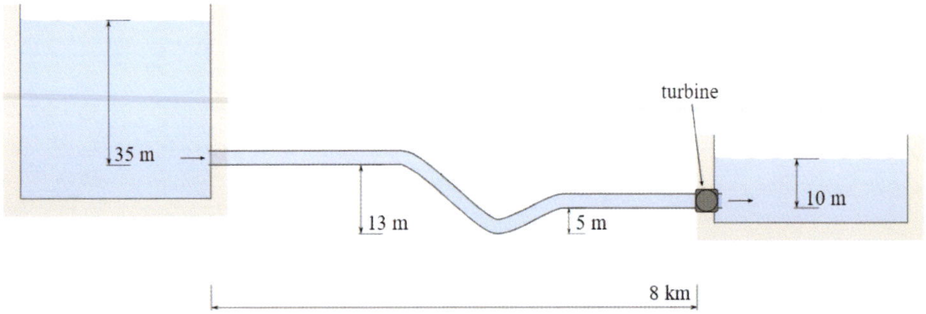


**Figure 58 – A pipeline between an upper and a lower reservoir**

The world's largest *"batteries"* are elevated reservoirs for grid-scale energy storage. Pipelines provide a path to and from a nearby lower-lying reservoir. We use electric pumps to raise water up, accumulating energy. Then, we let the water down through turbines, driving generators to release the potential energy back into the electrical grid. The efficiency of this overall system is 80%.

This sounds splendid; can't that serve as our energy storage? It can, but it's a matter of scale. A cubic meter of water weighs a metric ton ($\rho = 1\,000 \text{ kg m}^{-3}$). Raising its elevation ($z$) against gravity ($g$) does work ($W$) that we recover later:

$$W = \rho \cdot g \cdot \Delta z$$

So lifting that cube of water up by 200 meters stores 54 Watt-hours of energy. Britain's pumped hydro storage facilities hold 0.03 TWh of electricity. This is three orders of magnitude – *one thousand times* – smaller than the natural gas reserves accommodating seasonal fluctuations in energy demand. Strategic energy storage is a lifeline. It is the only way a country has to cope during deep winter, to heat homes and to fuel gas-fired peaker plants generating electricity.

Pumped hydro is a responsive energy storage method, but it is not a silver bullet.

# Applications of the circuit analogy

Those teaching electric circuits reach for an intuitive analogy with water pipes. A transistor works like a swing-gate which is triggered by a small minimum flow rate of water through a branch pipe, connected via a pulley to pull open a larger disc valve. This in turn allows a higher flow rate of water through the mainline.

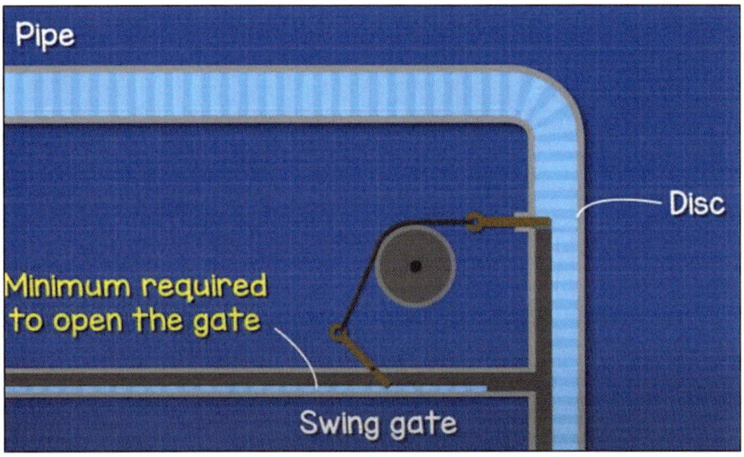


Figure 59 – How a transistor works, explained as pipes (Credit: TheEngineeringMindset.com)

Similarly, teachers of hydraulics reach for an analogy when a complex machine would be more familiar expressed in terms of more common electrical devices. A constriction in a pipe restricts flow rate like a resistor restricts flowing current.

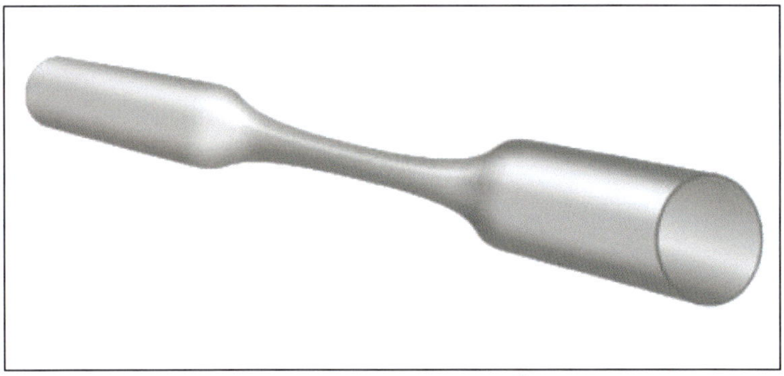

Figure 60 – A constriction in a pipe imposes a resistance against flow
Vanessaezekowitz / Wikimedia Commons / CC BY-SA 3.0

Mechanical engineers do incredible things with analogies, extending them to mechanisms. A hydraulic press uses a liquid in pipes of differing diameters to amplify a force, much as a lever on a pivot amplifies a force by mechanical advantage. The application of fluidics – pneumatics and hydraulics – for useful purposes is ancient. But human intuition is more at ease with a *lever* than with fluid devices. People always reached for familiar ways to intuitively understand *transport phenomena*, to see basic principles an engineered system abides by.

In the 1930's, Soviet architect and engineer *Vladimir Lukyanov* invented an analogue computer. It functioned by careful manipulation of water through a room full of interconnected pipes and pumps. The water level in various chambers represented stored numbers. Flow rates between them represented mathematical operations. This machine solved inhomogeneous differential equations: its successors were developed and used as recently as the 1980's!

**Figure 61 – The original programmer's development toolkit**

The circuit analogy can be helpful if we look both ways. We can glean valuable lessons about hydraulics from the world of electric circuits, and vice versa. Rates of change matter, at least in some applications, and equipment exists to handle them. Abstracting system components into distinct model items lets us model and describe them more easily than trying to build everything at once.

However, although abstracting mechanical systems to a circuit analogy has helped engineers devise innovative solutions in clever machines, We have not seen any such *"return on investment"* reported by anyone for fluid flow in pipes. The analogy is far removed, and may lead to a false sense of confidence.

Having the ability to see analogies between all kinds of domains is a skill that comes in useful, so long as it is approached with due care and attention. We must *not* insist that an analogy always exists perfectly, or indeed exists at all.

# Limitations of the circuit analogy

The caveat with analogies is that little bit of knowledge is a dangerous thing. Although the circuit analogy is, to some, a fascinating rabbit hole to dive down, it's not without its drawbacks. It risks teaching the wrong lessons to a budding fresh mind, curious about new concepts, if they jump to the wrong conclusions.

One big difference between steady state pipe flow and direct-current electrical circuits are that most liquid pipelines and all gas pipelines operate in a turbulent flow regime, in which pressure drop is more-or-less proportional to the square of the flow rate, and therefore behaves in a completely different manner compared to the electrical current through a resistor, where voltage drop is linear in the current.

| Domain | Linear Law | Nodal Rule |
|---|---|---|
| **Direct Current Electric Circuits** | Ohm's Law *(Universal)* | Kirchhoff's Current Law *(Universal)* |
| **Heat Transfer by Conduction** | Fourier's Law *(Universal)* | *(Rare)* |
| **Hydraulics of Flow in Pipes** | Poiseuille's Law *(Only some fluid flows)* | Flow Continuity *(Several forms)* |
| **Mechanical Systems** | Dashpot *(Not in every system)* | *(Rare)* |

**Table 2 – Domains that are in some ways analogous don't always abide by the same laws**

At best, we might get side-tracked from what *matters* in our domain. The nodal rule manifests as Kirchhoff's Current Law, so important to electrical engineers that they abbreviate it KCL. Flow continuity is important in liquids as well. But the thermal analogue, temperature compatibility, is almost unheard of because nobody needs it. Attempting to import techniques from one domain to address problems which don't exist in another domain would be an exercise in futility.

At worst, we might fall into the pitfall of assuming an analogy will exist at all. Ultimately, these systems are not completely alike. Although we can get a fair way building electric circuits to represent heat flows, and in fact people do that sometimes as it's cheaper than thermal experiments, this is only because the analogy between thermal conduction and electrical conduction through a solid happens to be strong. In fact, most good conductors of electricity are good conductors of heat! Even so, there is no thermal analogue for inductance. And we can't build circuits to represent the flow of fluids through our pipelines.

I have found a productive way to exploit an analogy is to look for places where it falls short. We gain insight not by insisting that a round hole must fit a square peg, but by *spotting the differences* and appreciating their wider implications.

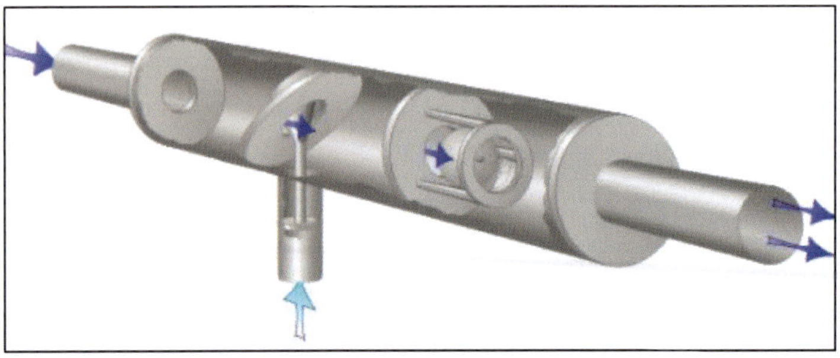

**Figure 62 – An electrical transistor is like a pressure-activated valve**
**Vanessaekowitz / Wikimedia Commons / CC BY-SA 3.0**

The world's largest machines, electrical power grids, transmit energy almost instantly. Turn on a device at home, and that power needs to have been generated somewhere at the speed of light. Water pipelines transmit transients very fast too, but even they can only transmit them at the speed of sound. Gas pipelines barely transmit anything in normal operation; they just pack and unpack, varying their inventory along the line – very unlike the electrical cables.

Flowing water experiences frictional losses at the roughness of pipe's inner surface, while moving charges are hindered at *all* points within a metal cable. They are akin to water being forced to flow through a pipe stuffed with sponge!

Though a check valve opens on reaching a minimum pressure drop across it, that's where the analogy ends. Diodes drop a fixed voltage across them (a flat 0.5 Volt, say) at all currents; check valves work nothing like that. Like any valve, a check valve imparts a pressure drop that varies depending on the flow rate. Moreover, the pressure drop across a check valve is typically fairly trivial on the scale of pressure drops in a pipeline system as a whole, whereas the same cannot be said about the voltage drop across a diode.

A liquid pipeliner closely monitors the actual volumetric flow rate through a pump, and the head it is imparting as shaft work, to avoid excessive cavitation. An analogous concept to cavitation is entirely absent in an electric circuit. An electrical resistor is typically linear, whereas flow coefficient is far from it. There are countless differences like this, and that's exactly where we learn.

# C4 – Shut-in pipelines and fluid properties

## What is shut-in?

Bottling up a fluid within any volume, such as the volume of a pipeline, is called *shut-in*. This is a common pipeline operation: valves are shut, so as to block any flow from entering, leaving, or traversing a station. The pipeline in this scenario can be thought of as akin to a pressure vessel. Surprisingly, fluids undergo change even in shut-in, as *operating conditions* vary over time.

## Pressure

*"We live at the bottom of an ocean of air"*. Looking down from a bird's eye view, an average person may estimate their cross section at 0.1 $m^2$. We are so deeply immersed in this ocean of air that it is equivalent to having us sit ten meters underwater. The air above me is applying pressure ($P$) with the force of a *ton* on my cross-sectional area!

The only reason the weight of the world on my shoulders doesn't cause me to implode is that my internal organs are also at this same pressure, pushing back in perfect equilibrium. But try snorkeling at any depth beyond just below the surface, and we are soon reminded why scuba divers are preoccupied with *pressure* – of their lungs, of their breathing apparatus, and of the deep water.

Pressure is not as intuitive as one first imagines. It is measured in units of energy per unit volume. A compressed air tank was long used in golf carts because as it decompresses, ($\mathrm{d}V$ with $V$ indicating the volume of air released and the "$\mathrm{d}$" prefix indicating a change, a notation we will see repeatedly in this book) it releases some of its energy $\mathrm{d}E = P \cdot \mathrm{d}V$ . This is a change in potential energy of compression, analogous to what physics teachers teach schoolchildren about a coiled spring ($k \cdot \mathrm{d}x$). It is, at the same time, the *compression work* due to a change in volume ($\mathrm{d}V$): conservation of energy dictates that a change in any potential energy is defined as *work done*.

Pressure is equivalently defined as force per unit area. A pressure drop applied to a pipe's cross-sectional area causes motion by accelerating a mass of fluid:

$$F = m \cdot a = P \cdot A$$

The point being made here is that pressure, energy and force are thoroughly intertwined; they are three concepts that are very closely related. A fluid flows:

- from higher pressure to lower pressure (in a simple horizontal pipeline)
- expending its accessible energy (the portion available for doing work)
- due to an imbalance of forces (acting from upstream and downstream)

The distinction between pressure, energy and force can be muddied if one is not careful, especially if non-specialists are careless when using imperial units.

> To remove water from a pipeline 3-ft in diameter, contractors pushed a foam pig along the pipeline using compressed air at 28 atmospheres, equivalent to 400 psig. They referred to this as *"four hundred pounds"*. The water exit line was only 1-ft in diameter, so they opened the end of the pig trap, placing a loader in front of it to catch the pig. The force was 250 tons, so the pig knocked over the loader and travelled 150 m (500 ft), destroying a wooden platform along its way! [29]

Those accustomed to imperial units might be tempted to refer to pressures in *"pounds"*, omitting to mention *"per square inch"* but knowing that it is implied. Some American pipeliners say *"pee-sigs"* (for psig) which seems clearer – the *"g"* referring to a gauge pressure, that is to say: above atmospheric pressure.

Whatever units we choose to work in, we look at a question in terms of different variables together – pressure, energy, force or head. Those working with fluids freely juggle these terms, finding any one to be most useful in precisely describing a fluid's behavior as it flows. A given problem will naturally lend itself to working with some subset, or all of these concepts at once. Whichever we choose, so long as we are careful, we are still looking at the same question, and so are bound to reach the same valid answer.

# Density, pressurization and compression

Saying a *compressible* fluid is affected by pressure sounds reasonable. But how does pressure affect a liquid which we are calling *almost-incompressible*? Well; not all pressurization is compression. When we squeeze something, if its volume contracts, we are *"compressing"* it. A volume per unit mass, a *"specific"* volume, is the inverse of density $(1/\rho)$. To compress is to change a *density.* *

But we can apply a pressure to a fluid without significantly affecting its volume or density! † Earlier we talked about squeezing a spring being like compressing a fluid; in an almost-incompressible fluid the *"spring"* is still there, it just doesn't move very much when we squeeze it. We can put just as much energy into the spring as we could put into a gas, and the spring still drives pipeline hydraulics – but it barely moves at all.

For this reason, in many pipeline calculations, we don't need to consider the density change caused by pressurization of, say, water at all. In those cases when we say water is incompressible, we mean it's *"effectively incompressible for our purposes here"*. We just *try* to squeeze, and it pushes right back at us, changing in density by a miniscule amount. To *pressurize* a fluid is to exert a force on it. Pressure after all is just a way of saying force per unit area. How much it actually *gets compressed* by is a different matter entirely. So we are *perfectly within our rights* to apply a pressure to an almost incompressible liquid. A pump is distinct from a compressor precisely because pressurization does not necessarily cause a lot of compression. When a fluid is *almost incompressible*, its density barely changes however much we pressurize it. Whatever force we apply as pressure over a cross section, it exhibits very little compression.

Water is very close to being *completely* incompressible. This makes sense; in water, molecules are basically touching their neighbors already. There isn't any way for them to pack themselves any closer together. Having satisfied ourselves that pressure doesn't affect density by much for such liquids, we must be cautious. Even for a truly incompressible liquid, the density ***is*** noticeably affected by the *temperature*. The density of water, for example, does vary with temperature, albeit only slightly. Just as *"hot air rises, cool air falls"*, the same can be said for hot water and cold water!

---

* Some use *"specific gravity"* (SG), which is *standard density* relative to a reference $(\rho_{\text{std}}/\rho_{\text{ref}})$; *"standard"* being at some set of reference conditions, and *"relative"* meaning with respect to a reference fluid, that being dry air for gases, or water for liquids. Some then define *specific weight* to be that relative standard density (the *"specific gravity"*) multiplied by the gravitational acceleration on Earth $(g)$. Either may be indicated by $\gamma$.

† In practice, we can apply pressure to water without affecting its density *very much*. True incompressibility would introduce infinity in some calculations.

## Compressibility and bulk modulus

Now we've cleared that up, what about the other extreme, a fully compressible fluid? This is a gas or vapor, exhibiting a marked rise in density if pressurized, and conversely a fall in density if depressurized. The same quantity of compressible fluid material occupies different volumes at different pressures.

Compressible liquids exhibit a sort of intermediate behavior. These naturally include liquefied petroleum gas (LPG) and natural gas liquids (NGLs). A fluid's *compressibility* describes how *easy* it is to change its density – and we shall define it precisely in this discussion.

We talk about a liquid's *bulk modulus* ($B$), the reciprocal of compressibility ($\kappa$):

$$B = \frac{1}{\kappa}$$

A bulk modulus describes how *difficult* it is to compress and change its density. There are two definitions to be aware of that we'll state as is for now: isothermal bulk modulus ($B_T$), and isentropic bulk modulus ($B_s$) – the latter also being called an adiabatic bulk modulus (isentropic means reversible and adiabatic). The isothermal bulk modulus describes compression occurring at constant temperature, while the adiabatic bulk modulus refers to compression occurring without any heat entering or leaving the substance being compressed. Bulk modulus is in pressure units as it relates to pressures; in the limit of $\mathrm{d}P \ll B$ the fractional density change caused by a pressure change of $B$ will be $\mathrm{d}P/B$.

---

> *"If water's bulk modulus is 20 000 bar, that means a 200 bar pressure change will cause a density (or volume) change of about 200 bar / 20 000 bar or 1%"*

Compressibility factors for many liquids are given in the American Petroleum Institute's *Manual of Petroleum Measurement Standards*. The *ARCO* formula calculates a crude oil's bulk modulus based on its density. The bulk moduli of liquids range from under 2 000 to over 20 000 bar. [25]

For a nearly incompressible liquid this is an input parameter given to the equation of state; bulk modulus is known to within around 5%, and stays constant within a pipeline's operating range. This means an equation of state can be used on that basis. Most liquids are adequately described by that *"bulk modulus equation of state"*, but compressible liquids are an exception.

For a compressible liquid, or a gas, the bulk modulus is a result derived *from* an equation of state, for which a different approach altogether is employed. Gas density is described by several widely-used equations of state, more complex than the familiar ideal gas laws taught in schools. Although for all gases the ideal gas law does hold quite accurately at low pressures. *

For an ideal gas, the isothermal bulk modulus is always equal to the pressure. As any fluid flows along a pipeline, its pressure and temperature change. For a gas, we usually express the isothermal compressibility ($\kappa_T$) – which is the inverse of the isothermal bulk modulus ($B_T$). That changes as density changes:

$$\kappa_T = \frac{1}{\rho} \cdot \left(\frac{\partial \rho}{\partial P}\right)_T$$

In this way, the compressibility ($\kappa$) of a fluid tells us the fractional change in density associated with an absolute change in pressure:

$$\frac{\mathrm{d}\rho}{\rho} = \kappa \cdot \mathrm{d}P$$

The compressibility of air at atmospheric pressure is $10^{-5}\ \mathrm{m^2/N}$ which is to say $1\ \mathrm{bar}^{-1}$. This is greater than the compressibility of water ($5 \times 10^{-10}\ \mathrm{m^2/N}$) by five orders of magnitude. In plainer terms, a 3% change in air density results from a 0.03 bar change in air pressure, but that same 3% change in the density of water requires a 600 bar change in pressure – many times higher than we would ever encounter in a pipeline! For this reason, water can generally be treated as being *almost incompressible*.

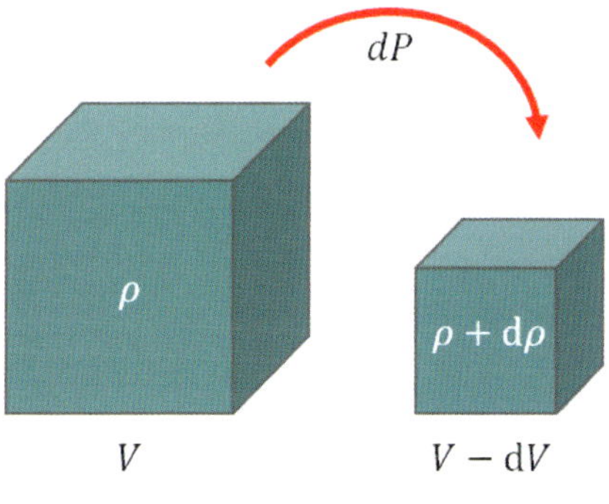


**Figure 63 – Compressibility ($\kappa$) is a change in density associated with a change in pressure**

* This is handy because it means that at standard reference conditions, the ideal gas law does apply, allowing us to do quick validation checks by relating standard volumetric flow rate to standard density in a gas pipeline.

# Temperature

Temperature ($T$) is a notoriously evasive property which becomes clear only if we define it carefully. Consider two blocks: one wooden, the other made of metal. If they have been left sitting in the same room all day, we know they are at the same identical temperature – namely that of their ambient environment, which is the air in the room. Despite this, they feel very different to the touch.

The metal object readily conducts heat, rapidly sucking away energy by heat transfer, striving to bring my hand's temperature into equilibrium with the cool room. It feels *"colder"* to the touch than the wooden object, which as an insulator, doesn't allow for much heat transfer. Touching something tells me more about its *thermal conductivity* ($\lambda$) than its temperature ($T$)!

Blocks of different materials at the same temperature feel very different to the touch. We can't sense temperature, what we feel is the material's conductivity to heat transfer.

We might choose to refer to the temperature of a fluid within a pipeline as the *line temperature*, to be clear we aren't referring to the temperature of its wider surroundings outside the pipe, which we might call the *ambient temperature*.

Even without considering time, the line temperature of a fluid changes along the length of a pipeline. A fluid is affected by how heat is conducted through the pipe wall material, surrounding layers of insulation, and the soil if it's buried. Heat might flow to or from the ambient air (or ambient sea, if it's underwater). In addition, passing through certain equipment, particularly regulators, pumps and compressors, significantly affects the line temperature. And the flow itself causes a change in temperature. Some pipelines install heaters or coolers to adjust the line temperature. A *thermal model* is the part of a pipeline simulator that handles the temperatures; we shall discuss it in its own dedicated chapter of the main book.

*The temperature profiles which arise in a liquid pipeline differs significantly to the temperature profiles for a gas pipeline. They differ both as we look axially, along the route of the pipeline, and also as we look radially, between the ground and the fluid.*

# Standard conditions and reference conditions

A pipeline, or indeed any industrial process, involves substantial variation in pressure and temperature. A set of reference conditions lets us speak about fluids and equipment *"as they would be at such conditions"*, independently of what the operating conditions happen to be right now at any given location.

Unfortunately, there is no universally accepted definition of what this is: it could be something close to ambient ($P_{\mathrm{amb}}$ , $15°\mathrm{C}$), or something sensible for the chosen system of units ($0°\mathrm{C}$ for SI, $100°\mathrm{F}$ for imperial). A given project or company should stipulate consistency across the board, but this is surprisingly challenging. Certain equipment, like flow meters, might come pre-calibrated against their manufacturer's reference conditions rather than those preferred by the pipeline operator. In many countries, pipeliners juggle several sets of reference conditions to serve different jurisdictions or meet customers' needs.

A set of reference conditions goes by several names. *Standard conditions* are standard pressure and temperature (STP), at which a given volume is called a *standard volume* ($V_{\mathrm{std}}$ in SI-units of $\mathrm{Sm3}$) etc. *Normal conditions* are another way of referring to the same concept: normal pressure and temperature (NTP) are those at which a given volume is a *normal volume* (units of $\mathrm{Nm3}$), etc. * They may also be referred to as *base conditions*.

Sometimes the reference conditions we choose can surprise us, like in the case of compressible liquids. Reference conditions for the liquid ethane in a liquid ethane pipeline must be defined at a high enough pressure for the ethane to actually be liquid. If it is a batched pipeline carrying batches of ethane alongside batches of other refined products, we might have to carry multiple different sets of reference conditions along with the batches! Gasoline batches in the same pipeline might well be at some legally stipulated reference conditions, conditions at which ethane would be gaseous. If we try to report, say, standard flows in two sets of reference conditions, we ought to be careful.

* The units of force (Newtons) are also denoted $\mathrm{N}$ by scientists. But pipeliners rarely use force, preferring to talk in terms of force per unit area: i.e. pressure.

# Enthalpy

Enthalpy ($h$) is a somewhat elusive quantity devised by cunning – though not *intentionally* cruel – thermodynamicists. There is a physical subtlety to enthalpy that many who use it might not fully appreciate. Its purpose is to capture the important interplay between two intertwined terms, a fluid's internal energy ($u$) and the work done in changing that fluid's pressure ($P$) or its density ($\rho$).

The work done by pressurizing, depressurizing, expanding or compressing a fluid (be that in a pump or compressor, or along a pipeline as it flows) is the change in pressure-times-volume ($P \cdot V$). Instead of talking about mass-specific and mole-specific volumes, we'll talk about densities, and state this as the pressure over density ($P/\rho$ i.e. $P \cdot \rho^{-1}$). This term tells us how much more energetic a fluid's internal energy would be if the operating conditions weren't hogging so much of its potential. * Some refer to this as pressure-volume work, or PV-work, while others choose to give it other names.

---

How much energy does it take to pressurize a pipeline?

Let's draw an imaginary box of a certain size, and call it a control volume. As a fluid parcel enters into or departs from our control volume as it flows, it conveys its own internal energy ($\mathrm{d}u$), which is a macroscopic measure of its molecules' bulk behavior. It also pushes the surrounding fluid out of its way to make space for itself. This being a genuine displacement, the parcel does work by virtue of its own motion ($\mathrm{d}(P \cdot \rho^{-1})$). This term looks strange to those who haven't seen it before, so let's explain how this equates to that work.

PV-work is done on the parcel as it enters a control volume, and it is done by the parcel as it leaves a control volume. The two cancel to zero only if pressure and density remain constant, which only holds true for an incompressible liquid. PV-work is considered distinct from the mechanical work applied externally via a pump or compressor's shaft, so we choose here to state it explicitly.

* There is a metaphor here somewhere for managing disgruntled employees!

# Head

A closely related concept is head, and it can be rather hairy.* It has units of energy per unit *weight*, or energy per unit *mass* – equivalent quantities related by dividing or multiplying by acceleration due to gravity ($g = 9.81\ \mathrm{m \cdot s^{-2}}$). This begs the question: *why?!*

Well; people talk in terms of head when looking at a liquid's hydraulic profile, and again for pump or compressor performance. Say we place a vertical tube open to atmosphere perpendicular to flow direction. The head ($H$) is defined as the height ($z$) of the surface of a liquid which a static tube would fill if it were attached to that point. It is dependent on the fluid's density ($\rho$), and is caused by the operating pressure ($P$) at that point.:

$$H = \frac{P}{\rho \cdot g}$$

Atmospheric pressure, for instance, pushes water in a *sealed* tube about ten meters high into the air. Any higher, and a vacuum forms at the top, causing water to turn into steam at room temperature. It sounds outrageous that we can vaporize water in this way, but we see it happen in clear tubing – it looks similar to boiling water, though the cause here is a vacuum not temperature. † Atmospheric pressure can be expressed as $10.3\ \mathrm{mH_2O}$. For an incompressible liquid, the head rise (length units) across a pump is:

$$\Delta H = \frac{\Delta P}{\rho \cdot g}$$

This is exactly the enthalpy ($\Delta h$) *added* by a pump if it has 100% efficiency. A real pump has an efficiency defined as the ratio of this ideal-pump enthalpy rise to true enthalpy rise, and the extra energy appears in the fluid as heat.

To express head rise equivalently in energy units ($\Delta H_{\mathrm{J/kg}}$ SI-units $\mathrm{J \cdot kg^{-1}}$) we multiply it by the acceleration due to gravity ($g$). This is the convention for gases. The above, for example, would have a value of head expressed in energy units of $\Delta P/\rho$ , but liquid pipeliners never do that.

* $h$ is a strong contender for physics' most overused symbol: enthalpy, height or hours.

† Turning water to vapor in this way is shown here (youtu.be/933XNdClFrc).
A puzzle for the reader: how can trees grow taller than this limit and still get liquid water from their roots all the way to the top?

# Hydrostatics: terrain affects pressure

Consider, if you will, an unmoving fluid sitting in a pipeline along a steep hill. Hydrostatics tells us it is at more pressure at the bottom of the slope than at the top. For a nearly incompressible liquid such as water, this calculation is a straightforward multiplication known to schoolchildren ($\rho \cdot g \cdot z$).

Now let's think about this: here we see a pressure drop, even though there is no flow. To say that *"pressure drop drives flow"* isn't *completely* false, but it's such a special case that it's best to forget it altogether.

---

What's the pressure difference across a 100 meter riser containing gas? What would it be for a liquid equivalent?

By using simulation, we can look at every point along the line as a *pressure profile*. As this line is stopped, this scenario shows us the *static pressure*.

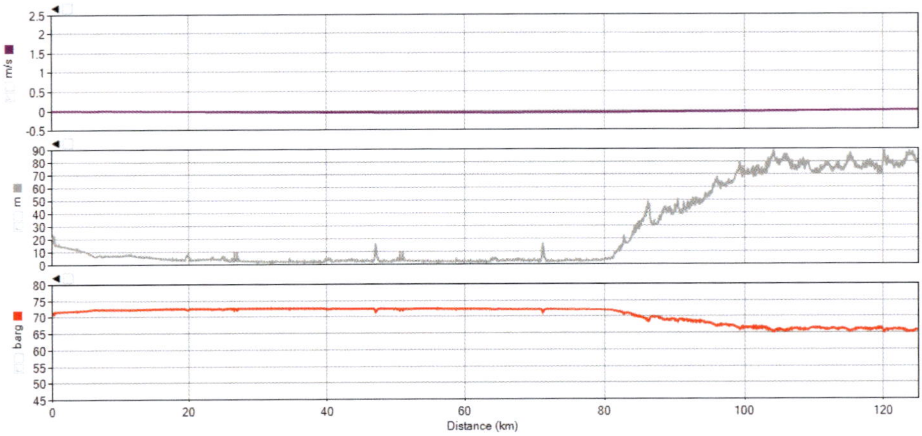


**Figure 64 – A stationary fluid (0 m/s) exerts *static* pressure, profiled here along a line's terrain. Because of elevation (grey), the pressure (red) is much lower at the top of a hill than at sea level.**

With a moment of contemplation we realize this pressure profile depends on *which* fluid happens to be in our pipeline. If the same pipeline contains the same volume of a heavier incompressible liquid (B) denser than a lighter incompressible liquid (A), even when that heavy fluid (B) sits still in shut-in, it exerts more pressure at the bottom of a given hill than does the light fluid (A).

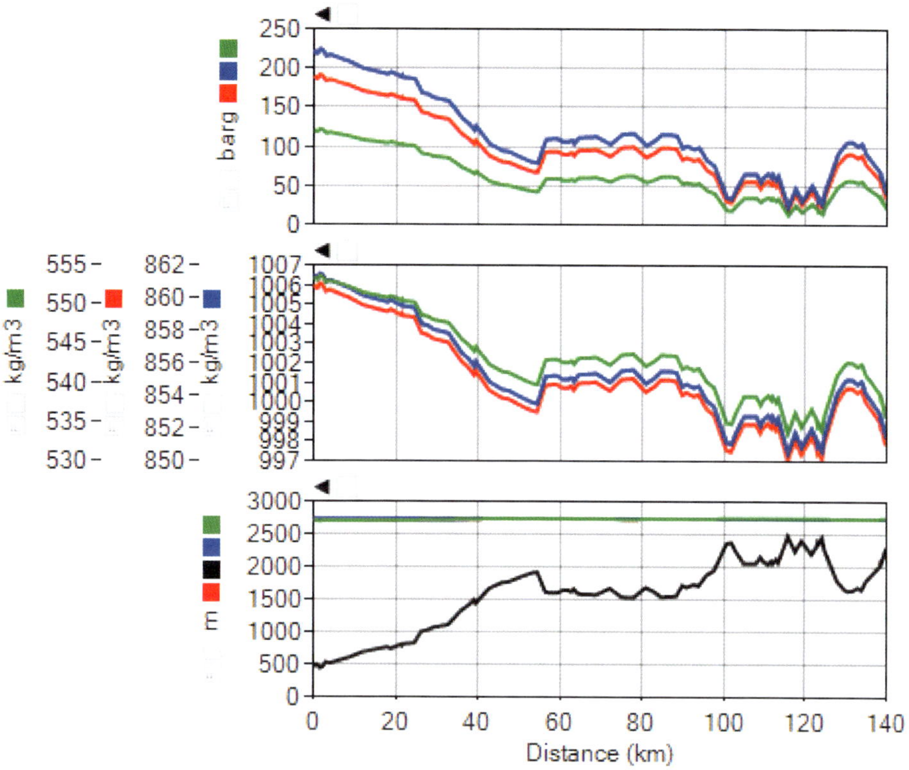


**Figure 65 – Diesel (red) is lighter than water (blue). Sitting still in the same pipeline, both exhibit the same head profile here (m), but pressure profile differs because they have distinct densities. Propane (green) is very compressible, so its density changes the most (5%) along the pipeline.**

In the case of a compressible liquid, or a gas, the fluid has a different density at every pressure and temperature, calculated by any good equation of state. The pressure at the bottom of this same hill is less obvious: density varies in a way that is not intuitively easy to predict, because compressibility factor varies.

---

Atmos SIM calculates the compressibility factor of a fluid using the appropriate equation of state.

# C5 – Fluids in motion and flow properties

## Hydraulics of pipe flow

The moment a fluid starts moving, we speak of hydraulics, a term that has several related but subtly distinct meanings. A civil engineer talks about water flowing over a weir. An automotive engineer is interested in thin tubes of fluid enabling mechanical advantage. A new reader delving into fluids is inundated with irrelevant works! But it doesn't have to be this way. The mysterious and overwhelming can just as well be made plain and obvious, with a little patient insight. Flow in pipes is well understood, and engineers use only a simplified set of tools and principles. *Pipeline* hydraulics, unsurprisingly, concerns only flow through a *pipeline*. We first examine, in detail, three vital – if approximate – statements about flow in a flat pipeline. Some authors may refer to these as the *pressure-throughput* relations:

- In a laminar liquid, flow rate is proportional to pressure drop
- In turbulent liquid, flow rate-squared is proportional to pressure drop
- In a gas, flow rate-squared is proportional to drop in pressure-squared

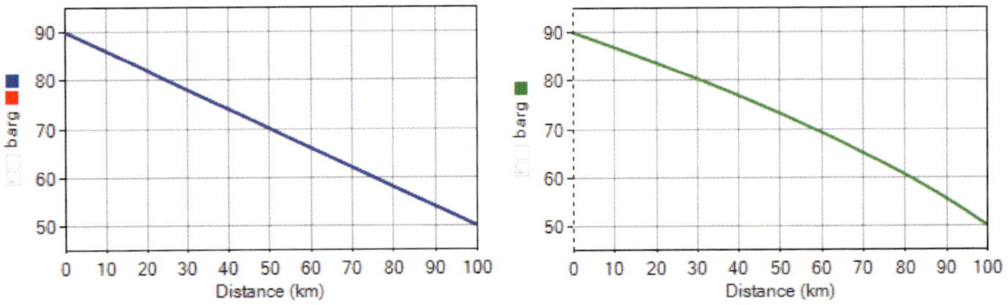


**Figure 66 – Liquids, in any laminar or turbulent regime, have a linear pressure profile (left) – if we neglect thermal effects. Gases flow in the turbulent regime and have a non-linear pressure profile (right). The flow is steady (unchanging) in all three scenarios along the same flat pipeline.**

A fluid's density and flow regime tell us how it's affected by friction. Along with the pipe's diameter and elevation, this lets us capture a snapshot of pressure and head throughout the pipeline, presented as profiles along a pipeline route like the examples above. We calculate how a fluid's energy manifests in its various forms along the route, deducing the flow rate for this pressure profile. At the heart of the *flow equations* quantifying these frictional losses is viscosity.

# Pressure drop is not always what drives flow

*"If I stand over here, and you are over there, can I drink your milkshake?"*

This is a playful way of asking if conditions at one end of a pipeline compel a fluid to flow towards the other end. When one asks the simple question: *"Why do fluids flow?"* the internet tells us that it is *"because they cannot support a shear stress when in static equilibrium".* This sentence, to me at least, doesn't help, and not just because I can barely understand what it means despite having a degree in the subject. Fortunately, like many of us, I was taught that flow is driven by pressure drop, which seems far more satisfactory. Unfortunately, this answer is quite misleading: *"for every complex problem there is an answer that is clear, simple, and wrong."*

A pressure gradient along a pipe can be expressed as a difference in pressure between any pair of points. It is this difference which we call the *pressure drop*. Pressure drop only scales linearly with flow rate in a flat pipeline containing a liquid which happens to be flowing in the laminar regime. Liquid pipelines typically operate from a few atmospheres up to 100 atmospheres or so, though as their infrastructure ages, operators will stick to more cautious pressures. In a gas pipeline across flat terrain, flow-squared scales with drop in pressure-*squared*. This is one reason gas transmission pipelines are operated at intense pressures, ranging from 15 to 200 atmospheres, some venturing even higher.

However, pressure isn't actually what drives anything. Rather, it is a kind of *total energy* which causes a fluid to flow. This is apparent from our previous thought experiment about fluids sitting still inside a hilly pipeline: hydrostatics dictates that our fluid is at a more intense pressure at the bottom of the valley. So there is already a pressure difference across a non-flat pipeline even as it sits still. Admitting this, we reach some important and profound truths. If a fluid flows down a hill, the direction of flow might well be *against* the pressure drop! In fact, pipelines heading downhill not only flow of their own accord, but they might even need to incorporate choke valves to *dissipate* that excess pressure.

# Head loss along a liquid pipeline

A liquid's pressure drop is affected chiefly by two factors: changes in *elevation*, and *friction* at its flow regime. A profile of head beautifully removes the complexities of elevation, showing us frictional losses. In a single-fluid pipeline, we plot our operating limits: *maximum allowable operating head* (MAOH) and *lowest allowable operating head* (LAOH). These are fixed profiles. Then, we try to *"shoot"* a linear head profile between them at any legitimate angle. This graphically finds throughput capacity from one station to the next, even with elevation changes.

When we talk about head along a pipeline, it is the height of those columns of fluid with respect to a baseline. So if we wrote the elevation of the top of the columns of fluid we would measure by tapping in with an open-topped tube, that would be the head. The pressure effects of a pipe going uphill or downhill get cancelled out. When we go from one batch to the next, we see a discontinuous jump in head, because the fluid in the column switches. This means that the head profile would not be a straight line.

**Figure 67 – A pipeline snakes its way down a mountain in Iceland**

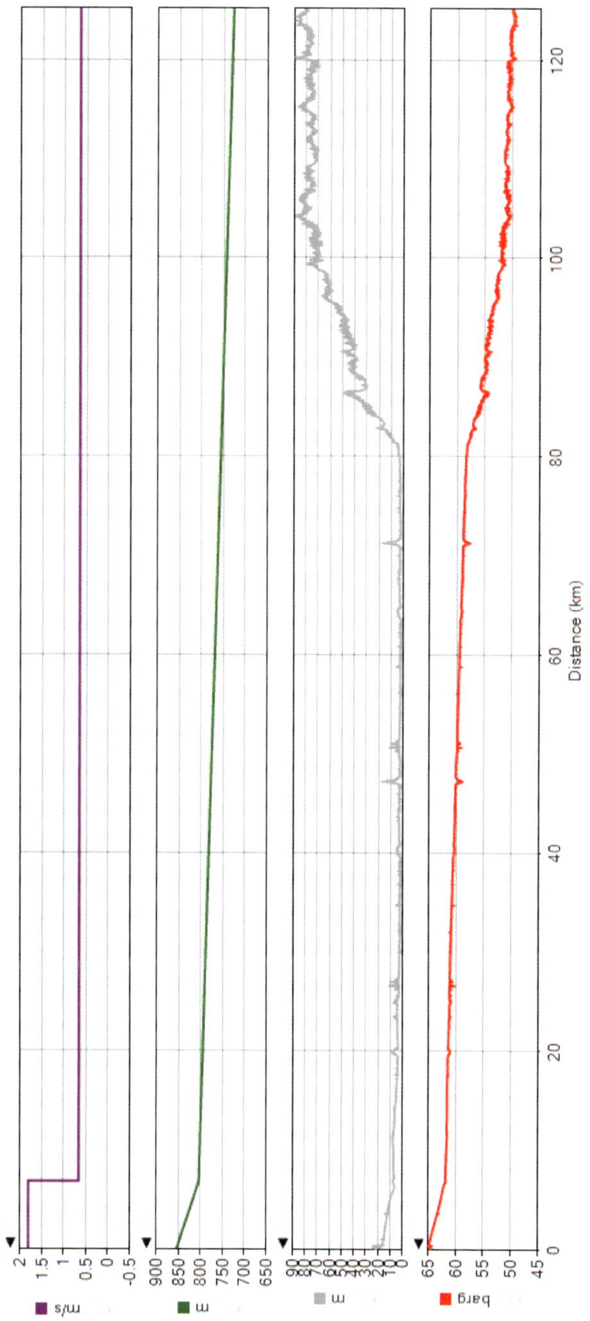


**Figure 68 – A flowing liquid being pumped uphill across a pipeline's terrain. It is at the same flow rate throughout, but a different velocity (m/s, purple), as diameter changes a few kilometers in. The pressure profile (barg, red) depends on elevation profile (m, grey), but the head profile (m, green) is straight through elevation changes, varying in slope only when velocity changes.**

This convenience is why instead of *pressure drop* across a liquid pipeline, hydraulic engineers talk about *head loss*. Moreover, a pump develops a fixed amount of head which gets *any* liquid over the subsequent hills, whereas pressure along that line depends on the fluid's density. Pressure will *change* depending on the batch. We see head in action if we build a transparent vertical pipe at the outlet of a pump. A column of liquid is raised by the pump, which in this case operates at zero-flow. Every fluid is raised to the same height, regardless of its density.

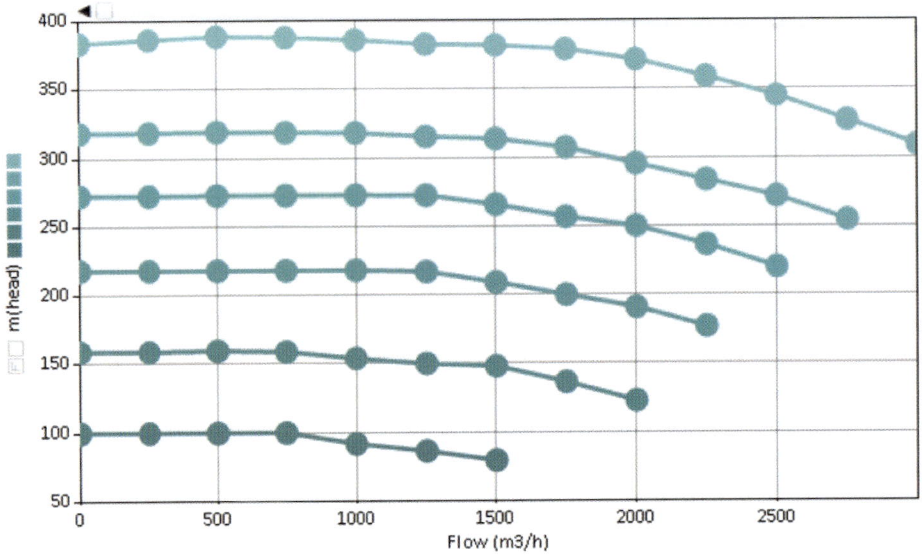


**Figure 69 – A map of head characteristic for a variable-speed pump in Atmos SIM**

What if the pump is operating at some flow rate, and we place the same vertical tube at *both* sides of the flowing pump? If our pump produces 100 feet of head, the vertical tube on the pump's discharge side rises 100 feet higher than the vertical tube on its suction side. In this case, we see *differential head* ($\Delta H$), so head at pump discharge is suction head plus the head produced by the pump.

Interestingly, the same applies to compressors. A centrifugal compressor imparts head in an almost completely analogous way to how a centrifugal pump does, so it is convenient to express it in units of *specific energy* ($\Delta H_{\mathrm{J/kg}}$), with SI-units of Joules per kilogram ($\mathrm{J \cdot kg^{-1}}$), a Joule being defined thus in the kilogram – meter – second system:

$$1\,\mathrm{J} = 1\,\mathrm{kg \cdot m^2 \cdot s^{-2}}$$

This means that energy-units head is equivalent to stating head in units of height ($H$) – which has SI-units of meters ($\mathrm{m}$). All we need to do is multiply by the acceleration due to gravity ($g = 9.81\ \mathrm{m \cdot s^{-3}}$). Gas pipelines hydraulic profiles do not show head because the density of a gas varies substantially, so it wouldn't make a straight line.

# Flavors of flow rate

## *All flow rates matter*

Flow rate is commonly referred to in rather vague terms, [*] usually *"throughput"* and sometimes *"capacity"* – the latter being notoriously ambiguous. It is also convenient in common parlance to say simply *flow* when intending *flow rate*.

More properly, flow rate can be described in several ways, any of which implies the rest given fluid composition and local conditions. That is to say: at a given point, and at a given instance, if one flow rate is known, fluid composition along with local operating conditions are enough to calculate the rest. *But* this does not mean that they are necessarily interchangeable. Far from it.

The conversion between a pair of flow rate-types *changes* with composition and with conditions. This means that there might be a unique conversion point-to-point *and* over time. Therefore, in a pipeline simulator, we might want to meaningfully assign *several flow constraints* within a scenario, as it is possible and potentially useful to *"switch"* from one flow constraint to another.

Just as we can use density ($\rho$) to convert between mass ($M$) and volume ($V$), or standard density ($\rho_{\text{std}}$) to convert between it and standard volume ($V_{\text{std}}$):

$$M = \rho \cdot V = \rho_{\text{std}} \cdot V_{\text{std}}$$

We can similarly freely convert between:

- mass flow rate ($Q_{\text{mass}}$)
- actual volumetric flow rate ($Q$), often called *"actual flow"*
- and standard volumetric flow rate ($Q_{\text{std}}$), often called *"standard flow"*

$$Q_{\text{mass}} = \rho \cdot Q = \rho_{\text{std}} \cdot Q_{\text{std}}$$

> A compressor processing high-pressure gas is generally *smaller* than one processing low-pressure gas, as the same mass of gas is denser at higher pressure, and so it occupies less volume.

[*] There are all manner of *"flows"* out there that don't concern us in the least. Powders and granulated materials flow, as do fluids in open channels, and fluids against plates. The internet and many books – popular and reference alike – seem intent on going all out, diverting our attention far from liquids and gases at pipeline conditions. This is surprising – pipelines are most interesting.

## *Velocity*

We frequently talk about a flow velocity ( $v$ in $\mathrm{m \cdot s^{-1}}$). By this we mean *axial* velocity, directed along the axis of the pipeline. But we also generally mean some sort of *average* velocity: in laminar flow the true fluid velocity varies across the pipe's radial cross section, and in turbulent flow it varies somewhat randomly over space and time due to turbulent vortices. In both those cases we perform the necessary averaging, so that our flow velocity ($v$) will refer to a velocity that is short-term time-averaged and also averaged across a pipe's internal cross-sectional area.

**Figure 70 – Velocity profile across a pipe's cross section in the laminar (left) and rough-pipe turbulent (right) regimes. We are only interested in *short-term time-averaged axial* velocity ($v$).**

Designers and operators keep a close eye on allowable velocity. Beyond a certain limit, flow becomes quite noisy, and could cause damage. In a liquid line, a typical maximum is around 2 m/s (7.2 km/h or 4.5 mph). In a gas line, velocity is an order of magnitude quicker, and at around 15 – 18 m/s (54 – 65 km/h or 34 – 40 mph) the flow hits limits of excessive friction and noise. These rules of thumb serve as a good sanity check.

## *Actual flow rate*

*Actual volumetric flow rate* or *actual flow* ($Q$ in SI-units of $\mathrm{m^3 \cdot s^{-1}}$) is the volumetric flow rate observed by a meter. It changes with prevailing operating conditions: for a given mass flow rate, a different density would mean a different actual volumetric flow rate, the latter being velocity multiplied by pipe cross-section ($\pi \cdot R^2$ or expressed in terms of diameter $¼ \cdot \pi \cdot D^2$). It is the actual flow that determines how equipment behaves, such as pumps, regulators and valves.

Actual flow rate is not universally equivalent to a flow of mass or of molecules; it is only conserved for an incompressible liquid flowing in steady state. For a compressible fluid, even one flowing at a steady state, the same clump of fluid at the far end of a pipeline is under less pressure. A compressible fluid such as gas therefore occupies *more* volume at that far end, so both the actual flow and the volume occupied by a given mass increase along the length of a gas pipeline.

## Standard flow rate

*Standard volumetric flow rate* or *standard flow* ($Q_{\text{std}}$ in SI-units of $\text{Sm}^3 \cdot \text{s}^{-1}$) is the volumetric flow rate we would observe if we were at defined standard conditions of pressure ($P_{\text{std}}$) and temperature ($T_{\text{std}}$). This about equals actual flow rate in an almost-incompressible liquid, but a gas or compressible fluid occupies far less volume at a pipeline's operating pressure than at reference pressure. Gas flow is typically quoted in units like *million standard cubic feet per day* (MMSCFD) in imperial units or SI-units of *standard cubic meters per day* (SCMD or $\text{Sm}^3/\text{d}$).

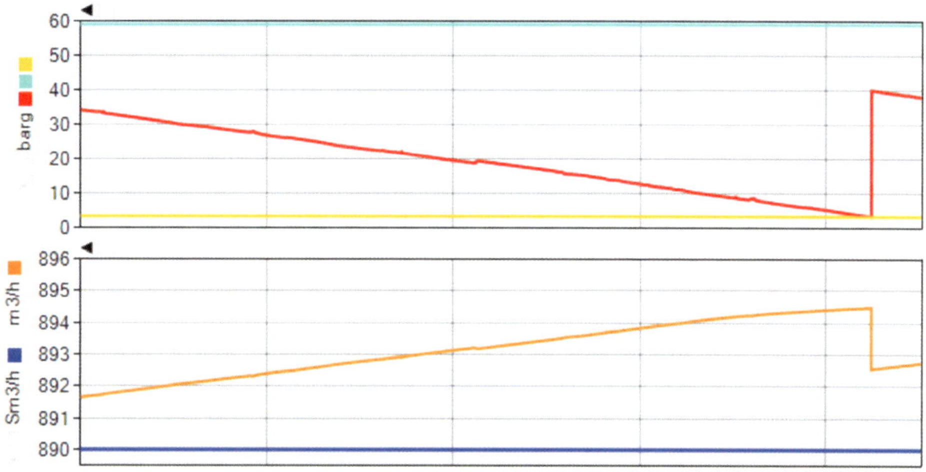


**Figure 71 – Actual volumetric flow rate (m³/h) and standard volumetric flow rate (Sm³/h) – commonly called *actual flow* and *standard flow* – differ along a compressible liquid pipeline**

*Standard volumetric flow rate* ($Q_{\text{std}}$) is equivalent to a *molar flow rate* for a gas, because at those standard reference conditions, the ideal gas law applies:

$$V = N \cdot \left(\frac{R \cdot T}{P}\right)$$

It's conserved even if composition varies, as a blend of ideal gases has the average of its constituents' densities, weighted by mole fraction. Moles are not conserved in reactions, but we don't model reactions. Blending of liquids is not so straightforward, particularly polar liquids. Liquefied petroleum gas (LPG) is routinely made from propane and butane by in-line blending at a mixing point. The blend has a standard density far less than the mean of its constituents, a phenomenon known as *shrinkage*. Mass is always conserved, and moles are conserved in mixing, but in this scenario, *standard volume* does not scale with the number of moles within our blend. Assuming that incurs errors of over 1%. To avoid this, for such a pipeline we would want to work with mass flow rate.

Standard volumetric flow rate is the flavor of flow rate of most interest to many pipeline operators because it is often nearly equivalent to a flow of money: a given standard volume of gas is a certain number of molecules and thus is worth a fixed amount. They think in terms of standard flow rate rather than in terms of flow velocity or actual flow rate. There are, however, some caveats to this – which we will come to shortly. One caveat is that gas pipeliners might also deal in terms of gross and net energy flow rates. It does not seem that any pipeliners trade in terms of mass flow rate.

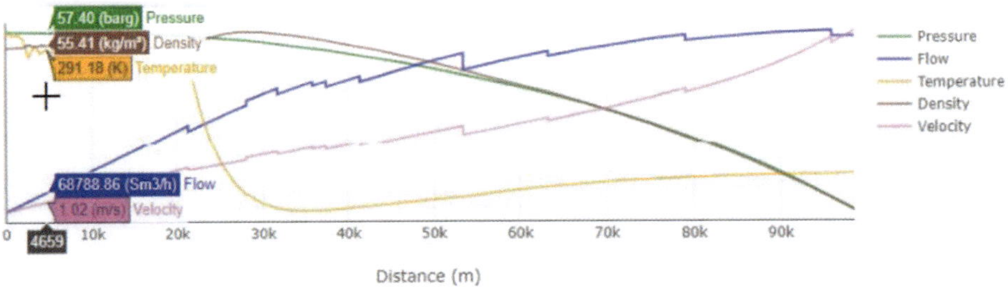


**Figure 72 – Standard volumetric flow rate is typically used on gas pipeline hydraulic profiles.**

## *Mass flow rate*

*Mass flow rate* ( $Q_{\mathrm{mass}}$ in SI-units of $\mathrm{kg} \cdot \mathrm{s}^{-1}$) is the only flow quantity that we can strictly say is always conserved, no matter what happens, in every conceivable scenario. * Changes in composition are subject to conservation of mass, even for polar liquids whose standard density does not distribute in an *"ideal"* fashion upon blending. Pipeliners rarely work in terms of mass flow rate, except in a handful of situations in which it is helpful to check conservation of mass rather than standard volume for validating inventory results.

## *Energy content as a flow rate*

Energy flow rates are the basis natural gas pipeliners charge their customers for energy, traditionally a higher *gross* calorific energy flow rate ($Q_{\mathrm{gross}}$). This has SI-units of Joules per second ($\mathrm{J} \cdot \mathrm{s}^{-1}$). † Because large quantities of energy are handled, pipeliners prefer units like terajoules per day ($\mathrm{TJ/d}$). Others define a *therm* in British thermal units (1 THM = 100 000 BTU), valued at about a US dollar at time of writing and is equivalent to about a hundred cubic feet of typical natural gas (1 CCF = 100 CF). The combustion of a fuel gas produces water, which inevitably *"wastes"* some of that

---

* It holds true in chemical reactions too; the only exception is nuclear reactions, and even there we talk about mass ($e = m \cdot c^2$) and never standard volume!

† This is the same as Watts, though few people seem to notice that.

gross energy in evaporating in the hot exhaust. This means that any consumer, including our own pipeline's compressions stations' fuel off-takes, can only actually make use a lower *net* calorific energy flow rate ($Q_{\mathrm{net}}$). Some gas suppliers now bill their customers on the basis of net calorific value. This allows for a range of gas compositions.

| Energy basis (gross calorific value) Natural gas / Hydrogen | Volume basis | |
|---|---|---|
| | Natural gas A / Hydrogen | Natural gas B / Hydrogen |
| **95 / 5%** | 85 / 15 vol% | 87 / 13 vol% |
| **90 / 10%** | 73 / 27 vol% | 76 / 24 vol% |
| **75 / 25%** | 48 / 52 vol% | 51 / 49 vol% |
| **60 / 40%** | 31 / 69 vol% | 34 / 66 vol% |
| **50 / 50%** | 23 / 77 vol% | 26 / 74 vol% |

Table 3 – The difference between defining a composition of hydrogen in rich natural gas (A) and leaner natural gas (B) on the basis of gross calorific value, versus based on volume [109]

An implication arises when hydrogen is blended into natural gas. If we say it's a 5% blend by gross calorific value, that equates in the case of a rich natural gas to it being a 15% blend by volume (denoted 15 vol%), but in the case of a lean natural gas it would be a 13 vol% blend. This has implications in general wherever natural gas composition is varying over time, as certain components have more energy value than others do.

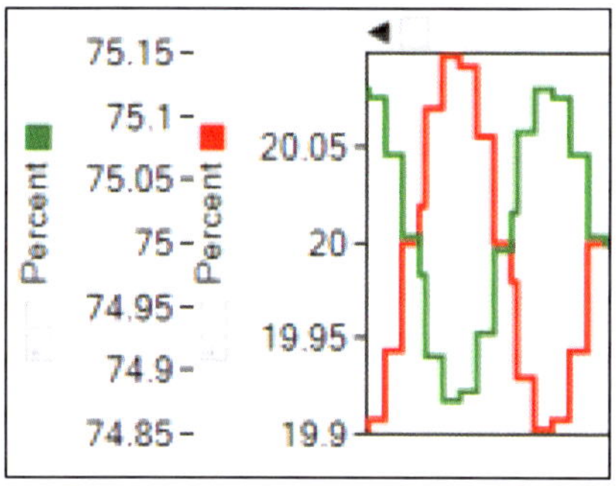


Figure 73 – The methane (green) and ethane (red) content of a natural gas varying over time

# Viscosity

## *Dynamic versus kinematic viscosity*

A viscous fluid is thick like honey. A cooler liquid becomes more viscous, just as warmer honey flows more easily. They say blood is thicker than water; well, to be more technical, we say "*cold blood is far more viscous than hot water*"! It is *viscosity* - not density – that describes if a liquid is thick or thin: it's harder to make viscous fluid flow.

Viscosity only has meaning in a fluid, manifesting as a resistance to shearing flows, as a *shear stress* ($\tau$). A shear stress is similar to a pressure: its units are force per unit area. But whereas a pressure pushes on the area, a shear stress describes a force parallel to the area. More shear stress arises in a more viscous fluid. Shear stresses arise when there is a shearing flow: adjacent layers moving parallel to one another at different velocities.

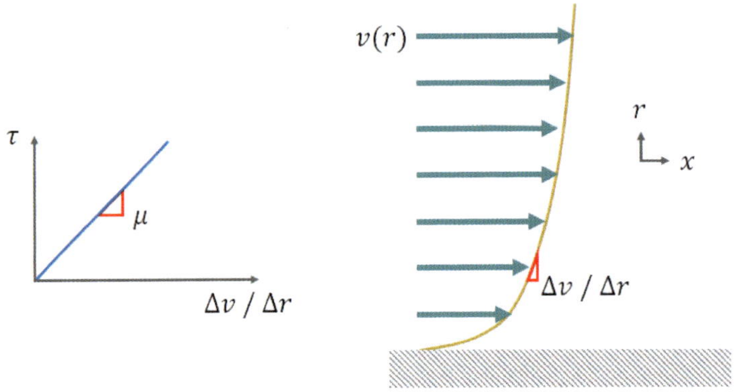


**Figure 74 – A Newtonian fluid's viscosity ($\mu$) relates shear stress ($\tau$) to velocity gradient ($\Delta v/\Delta r$)**

In a solid, a *strain* is defined as the amount the solid deforms per unit length; a *shear strain* ($\gamma$) is then the amount it's deformed in the direction normal to the length per unit length, i.e. because of a shear stress.

In a fluid flow, shear stress produces not just deformation, but *motion* – described by a *"rate of shear strain"* ($\Delta\gamma/\Delta t$), also known as the *"strain rate"*. In a Newtonian fluid, a category that includes most fluids moved through pipelines, the rate of shear strain induced by a shear stress is proportional to that stress. The constant of proportionality between the shear stress and rate of shear strain in a Newtonian fluid is known as the *dynamic viscosity* ($\mu$). In a fluid in laminar pipe flow, where the fluid velocity ($v$) is parallel to the pipe's axis, the rate of shear strain ($\Delta\gamma/\Delta t$) is equal to the slope of the radial profile of axial velocity ($\Delta v/\Delta r$). Shear stress ($\tau$) is proportional to the rate of shear strain ($\Delta\gamma/\Delta t$) via dynamic viscosity ($\mu$).

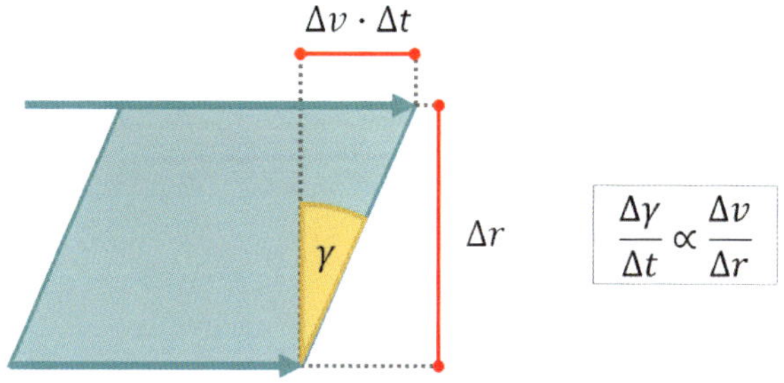


Figure 75 – A flowing fluid element deforms by a strain ($\gamma$): its top and bottom edges flow at different velocities. This strain changes at a rate proportional to the slope of velocity profile to the rate of shear strain, the velocity gradient ($\Delta v/\Delta r$).

In this way, the viscosity measures a fluid's resistance to flow in response to an externally applied force, and the frictional losses incurred along a pipeline. Dynamic viscosity is measured in several different equivalent units. 1 *centipoise* (cP) is defined as the viscosity of water at 20°C. This is a neater reference point than to say water's viscosity is a thousandth of an SI-unit, so the centipoise unit convention is very popular:

$$10^3\ \mathrm{cP} \equiv \mathrm{kg}\cdot\mathrm{m}\cdot\mathrm{s}^{-1} \equiv \mathrm{Pa}\cdot\mathrm{s} \equiv \mathrm{N}\cdot\mathrm{s}\cdot\mathrm{m}^{-2}$$

We find that dynamic viscosity ($\mu$) often appears in equations as a ratio with density ($\rho$). So we define dynamic viscosity divided by density as the *kinematic viscosity* ($\nu$). It is a fluid's inherent resistance to flow under its own weight:

$$\nu = \frac{\mu}{\rho}$$

The units of kinematic viscosity are the *centistoke* (cSt), where again 1 cSt is water's viscosity at 20°C. By that definition the units fall into place as follows:

$$10^6\ \mathrm{cSt} \equiv \mathrm{m}^2\cdot\mathrm{s}^{-1} \equiv \frac{\mathrm{kg}\cdot\mathrm{m}\cdot\mathrm{s}^{-1} \equiv \mathrm{Pa}\cdot\mathrm{s} \equiv \mathrm{N}\cdot\mathrm{s}\cdot\mathrm{m}^{-2} \equiv 10^3\ \mathrm{cP}}{\mathrm{kg}\cdot\mathrm{m}^{-3}}$$

## *Viscosity of liquids and gases*

The viscosity of a liquid is almost independent of *pressure*, so that effect is neglected. *Temperature* strongly affects the viscosity of liquid hydrocarbons. In a liquid, each molecule is in contact with its neighbors. As there's a radial velocity gradient $\mathrm{d}v/\mathrm{d}y$, molecules in adjacent streams drag on each other with intermolecular forces. Warmer temperatures mean a liquid's molecules have more kinetic energy of random motion, able to bounce past each other more easily in the presence of the same intermolecular forces opposing them.

The rate of thermally-induced state changes in a quantum mechanical system varies as the exponential of the energy involved in the state change divided by absolute temperature ($T$). For liquids molecules to flow past one another requires bending of molecules, so we expect this to follow a similar form, and indeed it does. The way a liquid's viscosity ($\mu$) depends on temperature for many liquids is well-described by the *Andrade* equation:

$$\mu = a \cdot \exp\left[\frac{b}{T}\right]$$

If we don't know the Andrade parameters ($a$ and $b$), we capture viscosity's temperature dependence using its value at two reference temperatures ($X, Y$):

$$\mu = \mu_X \cdot \left(\frac{\mu_Y}{\mu_X}\right)^{(T^{-1} - T_X^{-1})/(T_Y^{-1} - T_X^{-1})}$$

This equation can be extended to interpolate between multiple pairs of points. The Andrade form interpolating between a single pair of points might not hold true over large ranges of temperature, especially for heavy crude oils.

The processes refining crude oil into products converts sulfur to sulphonates. These are detergents whose presence strongly affects a liquid's viscosity. For this reason, the viscosity of refined products, especially diesel fuels, might vary significantly from textbook values. Even the viscosity of a given batch of crude oil might well be unknown to the operator. A shipper's stated value of viscosity might be off by a factor of several.

In a gas, like in a liquid, the viscosity has no pressure dependence. * Unlike in a liquid, a gas becomes *more* viscous when it is warmer. However, gas viscosity is so minuscule in the first place that this effect doesn't turn out to be the major influencer of head loss associated with a flow. So although there is an equation describing the dependence of the viscosity of a gas on temperature – Sutherland's equations – we can often assume to a good approximation that the viscosity of a gas is constant regardless of temperature.

* Here we mean *dynamic* viscosity ($\mu$). Kinematic viscosity ($\nu$) has a significant pressure dependence because the density of a gas depends on pressure.

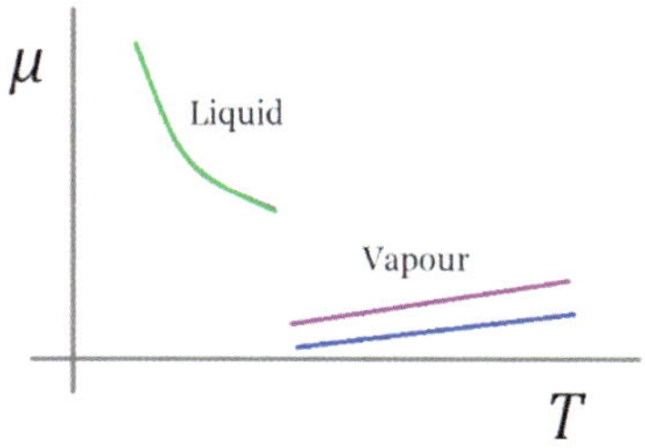


Figure 76 – A liquid's dynamic viscosity ($\mu$) *falls* with temperature ($T$) independently of pressure. Gas viscosity is also almost independent of pressure, but it *rises* with increasing temperature.

There is a physical reason gas viscosity increases with temperature. In a gas, molecules spend most of their time separated from their neighbors. Randomly directed thermal motions mean if there's a $\partial v / \partial y$, molecules from one stream wander into neighbouring streams. Those are moving faster or slower than the original stream, so as a molecule wanders it carries its own axial momentum along with it. Higher temperatures cause more random thermal motion, and accordingly more transfer of shear momentum. The kinetic energy of gas molecules' thermal motion is proportional to their temperature. This means that the average thermal velocities scale as the square-root of temperature, and so does the dynamic viscosity of an ideal gas.

| Fluid | Dynamic viscosity ($\mu$), centipoise | |
|---|---|---|
| | at 20°C | at 50°C |
| **Hydrogen** | 0.009 | 0.010 |
| **Methane** | 0.011 | 0.012 |
| **Air** | 0.018 | 0.020 |
| **Water** | 1 | 0.5 |
| **Milk** | 2 | 1 |
| **A heavy crude oil** | 42 | 19 |
| **A typical honey** | 15 000 | 1 000 |

Table 4 – Dynamic viscosities of liquids and gases at two different temperatures

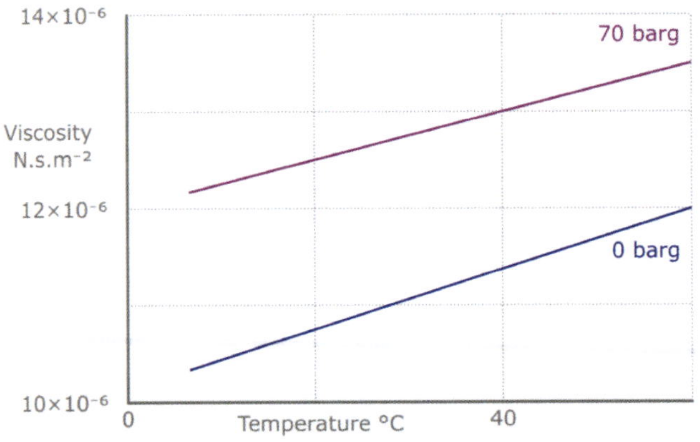


**Figure 77 – Methane, like all gases, becomes more viscous when warmer. The viscosity of a gas is minuscule, and varies remarkably little with pressure and temperature.**

## *Non-Newtonian fluids*

We have been assuming that the fluid's viscosity ($\mu$) is independent of the shear rate, and independent of the fluid's history. This is generally true for all gases, all refined liquid products, and most crude oils. We call those fluids *Newtonian*.

Non-Newtonian fluids are also pipelined. One example is slurries for the mining industry. We won't be covering those in this book, but there are known methods for modelling their properties, flow characteristics and behavior in a pipeline.

# C6 – HOW FLUIDS FLOW THROUGH PIPELINES

## Flow of liquids

### *Bernoulli's principle: kinetic, pressure and elevation head*

Flow of *incompressible liquids* is famously described by *Bernoulli's principle.* It is usually applied between two points (A and B) and expressed in units of *head* (in meters). It involves only fluid velocity ($v$), elevation ($z$), pressure ($P$), density ($\rho$), and acceleration due to gravity ($g$) in its ideal original form:

$$\frac{v_A^2}{2 \cdot g} + \frac{P_A}{\rho_A \cdot g} + z_A = \frac{v_B^2}{2 \cdot g} + \frac{P_B}{\rho_B \cdot g} + z_B$$

All of the terms in this equation can be relatively important in various scenarios.

Let's work an example. Consider a tank of water ($\rho = 1\ 000\ \mathrm{kg/m^3}$) with a hole in its wall one meter below the liquid surface. How fast does water emerge from the hole? Applying Bernoulli's principle along a path from the water's surface (A) to the outside of the hole (B):

$$P_A = 1\ \mathrm{atm}, v_A = 0, z_A = 0$$

$$P_B = 1\ \mathrm{atm}, v\ = ?, z = -1\ \mathrm{m}$$

Substituting those in we find that:

$$½ \cdot v^2 = -g \cdot z\ so\ v = \sqrt{2 \times 9.81\ \mathrm{m \cdot s^{-2}} \times 1\ \mathrm{m}}$$

$$= 4.4\ \mathrm{m/s}$$

What just happened here? Are we conserving energy along this path, and just moving some gravitational potential energy to kinetic energy? We are not! To see that something more complicated must be going on, consider the point just inside the tank at the hole (C). At that point:

$$v_C = 0$$

$$z_C = -1\ \mathrm{m}$$

How come? The answer is that there is a pressure term in Bernoulli's equation, which represents the capacity to do work. Inside the hole there is more pressure; as we trace a path down from the surface of the water to just inside the hole, our notional fluid element is *"doing work"* to place itself in a higher-pressure environment, and then when it goes sideways out of the hole, *"work is done on the notional fluid element"* to near-instantly accelerate it to 4.4 m/s.

So Bernoulli's principle doesn't just conserve kinetic and potential energy. A third term says how much *"pressure work"* a fluid can do on its surroundings.

The basic form of Bernoulli's principle attained celebrity because it *is* important. It's a handy expression for water flowing over short distances, say through a nozzle or a leak. But it doesn't apply universally. It assumes a *steady* flow of *incompressible* liquid. We must always remain cognizant of these limitations. *

Most importantly: Bernoulli's principle was formulated to be used only in circumstances where frictional losses are negligible, as it does not account for that. We can add a head loss ($\Delta H$) to the right hand side of Bernoulli's equation:

$$\frac{v_A^2}{2 \cdot g} + \frac{P_A}{\rho_A \cdot g} + z_A = \frac{v_B^2}{2 \cdot g} + \frac{P_B}{\rho_B \cdot g} + z_B + \Delta H$$

For fittings, there are look-up values and rules of thumb for this head loss. But in pipelines, head is mainly lost to friction along the long route of the pipeline, not fittings. What do we do if there are significant frictional losses to consider?

## *Darcy-Weisbach equation: head loss due to friction*

The Darcy-Weisbach equation defines an empirical friction factor ($f$). Its premise is that head loss ($\Delta H$) along a pipe is proportional to its length ($L$) and to half the velocity-squared ($\frac{1}{2} \cdot v^2$), and inversely proportional to diameter ($D$):

$$\Delta H = f \cdot \frac{L}{D} \cdot \frac{v^2}{2 \cdot g}$$

How do we use this equation to calculate head loss *from* friction factor? In the laminar regime, friction factor ($f$) is calculated directly. But in the other regimes we are forced to deal with correlations to *"look it up"*. What is a regime exactly? That will be the topic of the next section.

---

* Other forms of equation are occasionally seen to express Bernoulli's principle by accounting for *compressible fluids* and even *transient flows*.

# Reynold's number and regimes of flowing fluids

*"Big whorls have little whorls that feed on their velocity;*
*Little whorls have lesser whorls, and so on to viscosity."*
– Lewis Fry Richardson, 1922

## *Inertial and viscous forces*

The particles of a fluid passing a cross-sectional slice as they proceed down the pipe are generally not all moving at the same velocity. In a pipe, this is because the fluid at the wall is stationary (zero-velocity). This no-slip condition creates a boundary layer at the edge of the flow. This fluid is subject to both *inertial* and *viscous* forces. Authors either state this without qualification, or elaborate it in confounding detail. So it is worth explaining what's going on. As we apply a greater axial pressure gradient, it takes a larger radial gradient of axial velocity to produce a balancing shear stress in steady flow. This means peak velocity (seen in the center of the pipe) grows, as does average velocity.

As a thought experiment, let's start with a laminar flow. As we apply a greater pressure gradient, so the average velocity grows. Viscous forces tend to keep the flow axial, but if the fluid starts tumbling inertia tends to keep it tumbling.

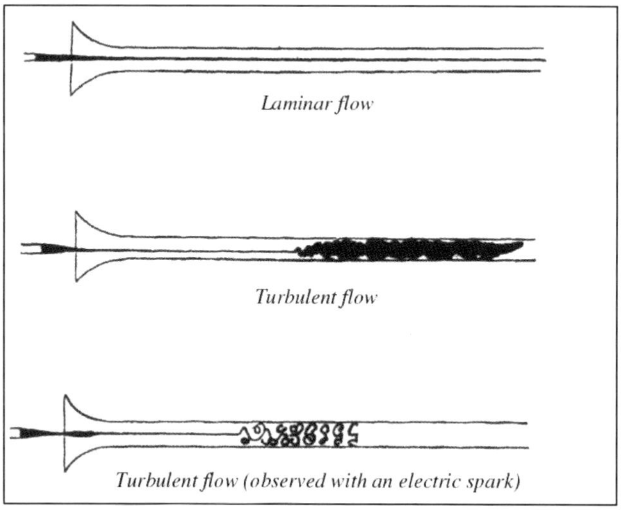


**Figure 78 – In 1883 Reynolds published these drawings.** ***"An experimental investigation of the circumstances which determine whether the motion of water in parallel channels shall be direct or sinuous, and of the law of resistance in parallel channels".***

In a turbulent flow, *random* motion is superimposed on the fluid's bulk forward motion. Fluid can't move through other fluid, so the random motion manifests as *vortices*. If every fluid element moves along closed loops, there are no collisions: vortices can be no bigger than the pipe. Nor, as we see below, can they be smaller than a certain limiting minimum vortex size.

Flow is characterized based on its ratio of inertial to viscous forces. This is expressed as a dimensionless quantity known as *Reynolds* number ($\mathrm{Re}$), calculated from fluid *density* ($\rho$), *velocity* ($v$) and *viscosity* ($\mu$) with reference to a length scale, which for flow in cylindrical pipes is the *inner diameter* ($D$):

$$\mathrm{Re} = \frac{\rho \cdot v \cdot D}{\mu}$$

The velocity times cross-sectional area ($A$) is actual volumetric flow rate ($Q$). If we choose to use *kinematic viscosity* ($\nu = \mu/\rho$) * we get the following form: †

$$\mathrm{Re} = \frac{Q \cdot D}{\nu \cdot A}$$

Reynolds number quantitatively measures whether viscous forces suppress small rotational motions, or whether those motions will grow. In this way, it informs us about the present *flow regime*. Experts see several shades between two extreme flow regimes: laminar and rough-pipe turbulent. Naming regimes is hopelessly inconsistent in reference literature, so let us label them A to D:

A. Laminar (Re< 1 000 or perhaps < 2 000)
B. Uncertain zone between laminar and turbulent (1 000 < Re < 4 000)
C. Smooth-pipe turbulent (4 000 < Re < 20 000)
D. Rough-pipe turbulent (Re > 100 000)

Between Reynolds numbers of 20 000 and 100 000, the flow regime could either be in smooth-pipe turbulence or in rough-pipe turbulence, depending on the pipe wall's relative roughness ($\varepsilon/D$).

---

* As the dynamic viscosity ($\mu$) acts in tandem with density ($\rho$), it is common to think in terms of kinematic viscosity ($\nu = \mu/\rho$), which is represented by the Greek letter *"nu"* and should not be confused for the symbol for velocity ($v$).

† Pipeliners may find Reynolds number using the standard flow rate ($Q_{\mathrm{std}}$). Diameter is about the same, but standard velocity ($v_{\mathrm{std}}$) differs from operating velocity and standard viscosity ($\nu_{\mathrm{std}}$) differs to viscosity at operating conditions.

$$\mathrm{Re} = \frac{Q \cdot D}{\nu \cdot A} = \frac{Q_{\mathrm{std}} \cdot D}{\nu_{\mathrm{std}} \cdot A}$$

## A – The laminar flow regime

Flow is predominantly *laminar* at very low Reynolds numbers (< 1 000 though some draw the limit at 2 000 and it seems a little arbitrary). Laminar flow is a function of pipe diameter, and of the fluid's velocity and viscosity. It is orderly, with parallel streamlines without mixing. It occurs when the ratio of inertial to viscous forces is low; that's when the fluid is thick, and the flow is slow.

I always found this business of flow regimes incredibly dull, and thought that those teaching the subject overcomplicated matters. It's all really quite intuitive. Lamina through a pipe have a sort of telescopic effect. I recently assembled a swivel chair, and laminar flows are a lot like that. LAMINA is just Latin for plate, and in pipeline fluids we use it to mean a *concentric hollow cylinder*. The outermost is stationary because it's in contact with the wall. Then, each plate (each lamina) slides past the next, until the innermost glides freely.

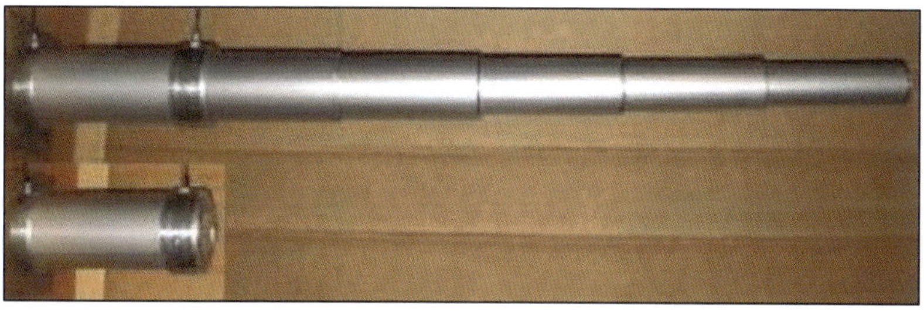

**Figure 79 – Just as the middle layers of a telescopic cylinder are free to extend or retract, the layers (laminae) near the middle of a laminar flow are unconstrained by the wall. The non-slip condition only keeps the *first* very thin layer (lamina) stopped at the wall.**

It caught me out, as the inner cylinder of the telescope fell to the ground, and I reeled back surprised. The chair was now in more pieces than I started with!

For this reason, a laminar flow is not affected by roughness. The roughness of a pipe wall's inner surface only interacts with the layer touching it, the outer lamina, and that outer lamina is stopped. That means that however rough a pipe wall may be, it doesn't matter. Even if the roughness extends through several of our notional lamina, those too are nearly stopped. So in a laminar flow regime, the important shear stresses act between *adjacent* lamina, and so the friction that flow experiences is not influenced by the wall's roughness.

In this laminar flow regime, fluid moves smoothly, like honey squeezed from a bottle. Each concentric cylindrical layer slides against the next unimpeded; without resistance. Each layer is infinitesimal, so in the laminar flow regime, *friction factor* is independent of any typical pipe roughness. Roughness acts only in a part of the fluid that is moving extremely slowly. By contrast, adjacent lamina are moving more briskly relative to one another, and that happens all the way through the pipe, not just in this thin region at the very edge.

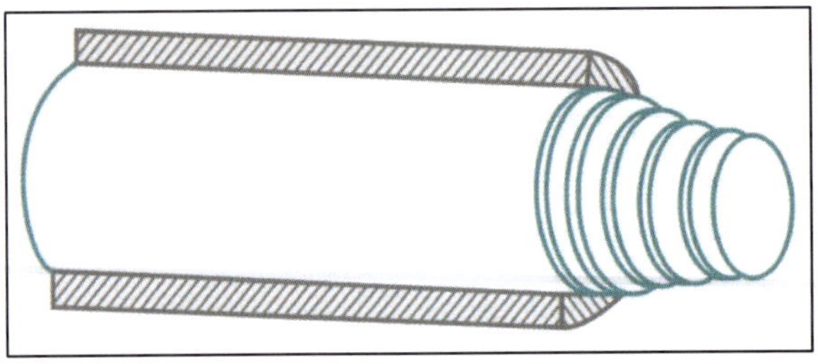

Figure 80 – In a laminar flow through a pipe, fluid layers move in concentric cylinders. This is how heavy crude flows slowly through small pipelines.

The lamina slide independently past one another as the effect of inertia is negligible. In pipelining, the only fluid seen routinely operating in the laminar flow regime is very *heavy* crude oil flowing very *slowly* through *small* pipes. If we are very careful we can force laminar flow even at Reynolds numbers as high as 80 000. But a real pipeline probably enters a turbulent regime by the time it reaches a Reynolds number of a couple of thousand; certainly by 5 000.

A gas flowing through a pipeline is essentially never in the laminar regime.

---

For roughness to have no effect in an 8-inch pipe:

- how slowly must crude oil flow?
- how slowly must water flow?

There are almost no Navier-Stokes equations that are solvable with pencil and paper. Here we get to solve one of the only ones! The *Hagen-Poiseuille* derivation gives us an analytical expression for friction factor in laminar flow. By definition as a universal truth with no need for any empirical correlations, the *Darcy* friction factor ($f$) is inversely proportional to Reynolds number ($\mathrm{Re}$):

$$f = \frac{64 \cdot \mu}{\rho \cdot v \cdot D} = \frac{64}{\mathrm{Re}}$$

## B – The uncertain zone of laminar-turbulent transition

Around 1 000 < Re < 4 000 a poorly characterized *zone of uncertainty* surrounds the threshold between laminar and smooth-pipe turbulent flow. In this regime, flow might be laminar, or it might be smooth-pipe turbulent: we can't easily predict which of the two very different behaviors it will exhibit. * What matters here is that it is poorly characterized. As we can't confidently describe which flow regime will arise within this range of conditions, this range of Reynolds numbers is best avoided when pipelining.

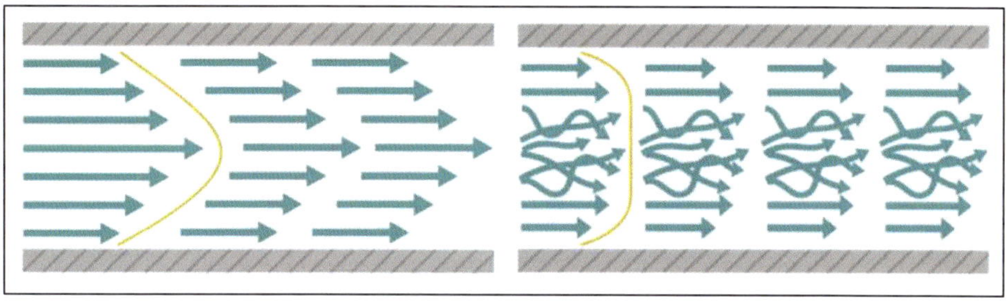

**Figure 81 – There is a zone of uncertainty between well-defined flow regimes, within which we can't be sure if the flow regime is laminar (left) or smooth-pipe turbulent (right)**

In this uncertain zone, friction factor ($f$) changes drastically in a discontinuous fashion. How do we model this? Solvers of partial differential equations (PDEs) can't handle such non-linearity in the equations we throw at them. Normally we mitigate non-linearity by making tighter spatial knots or taking shorter time-steps. But here, a discontinuity is bound to be terribly non-linear, no matter how fine our time and space meshes may be!

We don't know what disturbances are happening. How rough is the surface of a 20-year-old pipe? Do bends or fittings cause small vortices? Do acoustic disturbances happen outside the pipe? We won't get perfect accuracy, so we aim for good behavior that won't cause a simulator to fall over. A cubic spline interpolation from laminar at a Re of around 2 000 to turbulent at 3 000 or 4 000 is both continuous and continuous in its first derivative with respects to Re. Errors introduced here are short-lived on most systems, as pipelines only operate in this region briefly during start-up or shut-down.

* Some call this a *"critical zone"* – it seems engineers have taken a liking to calling anything of notable interest *"critical"*.

## *C and D – The turbulent regimes and the viscous sublayer*

Turbulent flows are complex and poorly understood even today, so these are empirical formulas. In laminar flow, *viscous forces* transfer axial momentum back and forth across a pipe. In turbulent flow, it is *vortices* that transfer axial momentum across a pipe's cross-section, and they do so far more effectively.

A vortex experiences an effective Reynolds number ($\mathrm{Re}$) whose characteristic scale length is the diameter of the *vortex* ($d$) not the diameter of the pipe ($D$):

$$\mathrm{Re} = \frac{\rho \cdot v \cdot d}{\mu}$$

For a vortex smaller than some size, Reynolds number is lower than 2 000 so viscous forces dominate inertial forces, so it can't form or won't be sustained.

As flow can't penetrate the pipe wall, vortices can't exist closer to the wall than their own vortex diameter. So in very close proximity to the wall, typically within a few dozen microns, flow is always laminar. We observe this to happen. The bulk time-averaged axial velocity is essentially constant across the pipe, except in a thin layer near the wall, called the *viscous sublayer*.

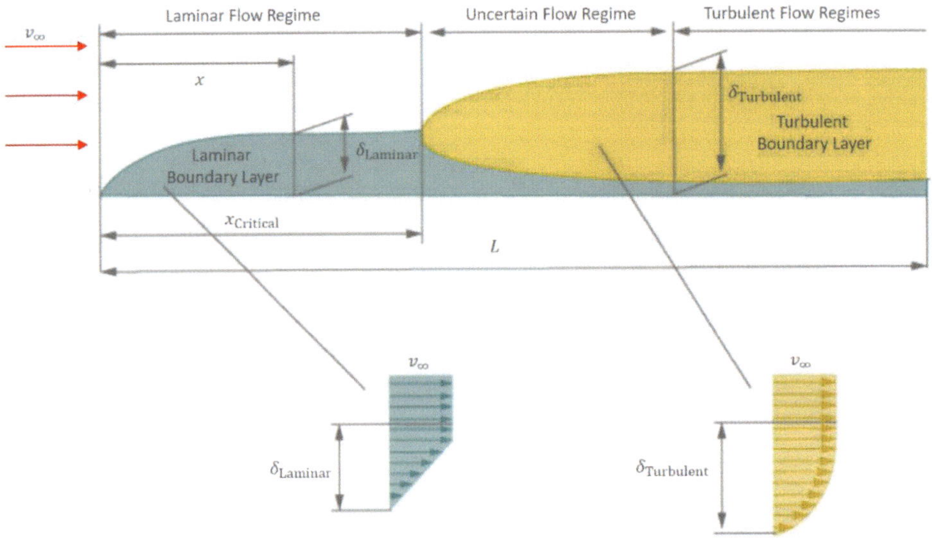


**Figure 82 – There is a laminar flow regime, an uncertain flow regime, and several distinct turbulent flow regimes.**

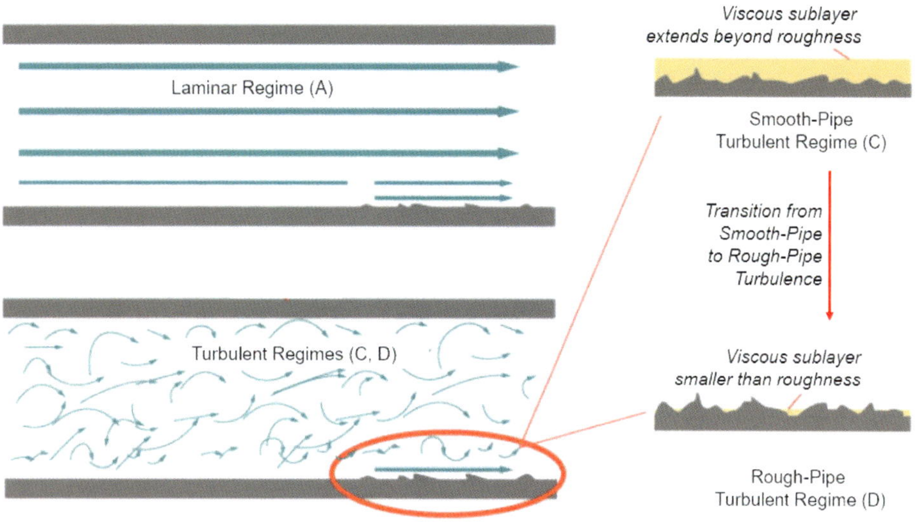


**Figure 83 – The laminar flow, smooth-turbulent and rough-turbulent regimes in a pipe. Three turbulent regimes' wall roughness extend out of the viscous sublayer, or not?** [26]

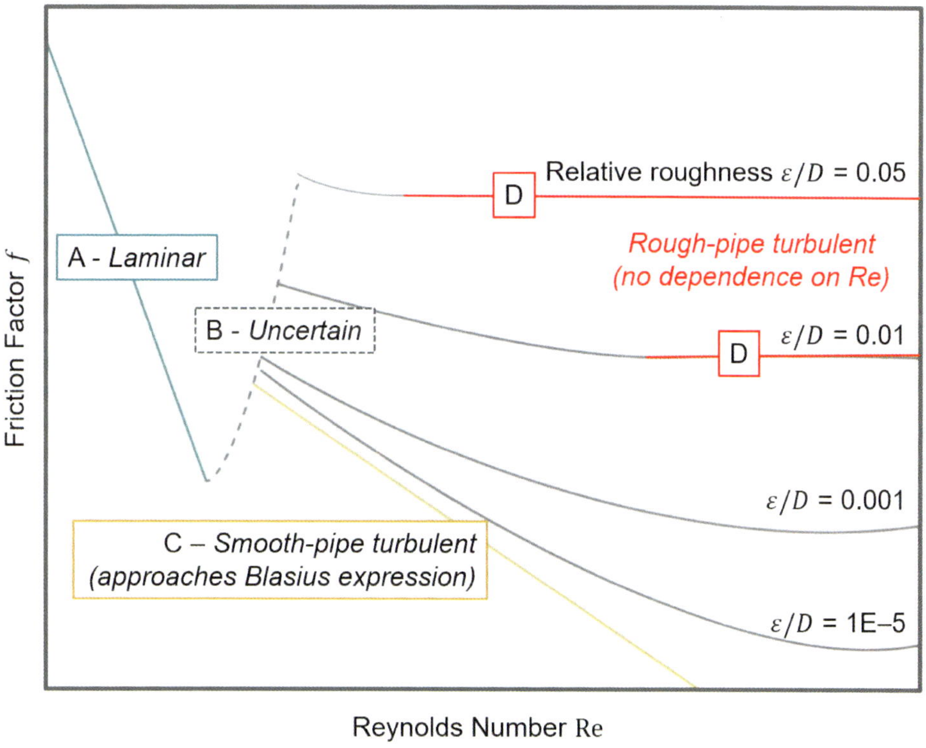


**Figure 84 – The laminar regime (A), uncertain zone (B), smooth-pipe (C) and rough-pipe (D) turbulence, on a Moody diagram of friction factor ($f$) versus Reynolds number (Re).**

## *C – Smooth-pipe turbulence*

A *smooth-pipe turbulent* flow regime is seen at moderate Reynolds numbers (4 000 to 20 000), in which both inertial and viscous effects act significantly. The center-line of the pipe is turbulent, and a viscous sublayer extends beyond the roughness of the pipe wall by a significant distance. Locally, a turbulent flow might involve a clump of fluid moving *backwards*. We think of the overall flow's velocity as having an *average magnitude* in the axial direction, along the pipe. This is sometimes called the smooth-pipe flow regime, because the wall's roughness is *"hidden"* inside the viscous sublayer, and friction is dominated by momentum transfer in the turbulent core.

This *"Prandtl form"* is a straight-forward friction factor ($f$) correlation which is applicable at all Reynolds numbers ($\mathrm{Re}$) in this smooth-pipe flow regime:

$$\frac{1}{\sqrt{f}} = 2 \cdot \log_{10}\left[\mathrm{Re} \cdot \sqrt{f}\right] - 0.8$$

We recommend using the Colebrook-White equation or GERG equation presented in the next sections, in the interests of having a single equation for turbulent friction factor in all regimes. These are close to the Prandtl equation in the smooth-pipe limit.

## *D – Rough-pipe turbulence*

At very high Reynolds numbers ($\mathrm{Re} > 100\,000$), there's a rough-pipe turbulent regime. It is utterly chaotic, with pressure fluctuations throughout. Here, the size of the pipe's wall roughness protrudes through the thin viscous sub-layer and disturbs the bulk fluid, affecting flow resistance and incurring pressure drop. Interestingly, friction no longer varies with Reynolds number. Vortices slam into the wall's roughness. They are good at laminar transfer of axial momentum, thereby efficiently transferring that momentum from the flowing fluid to the wall.

Ask for a universally agreed upon definition of turbulent flow and the response is *"you know it when you see it!"* Its characteristics are that its specifics are a mélange of sensitive, unpredictable chaos: it is intensely dependent on initial conditions. A slight change in any minor facet of its starting state completely changes how it behaves. All anybody can strictly claim to do with turbulent flow is to make "statistically likely" predictions describing it.

Nikuradse artificially roughened his pipes by gluing on sand. Based on his experiments, the following correlation was formulated to describe rough pipes:

$$\frac{1}{\sqrt{f}} = 2 \cdot \log_{10}\left[\frac{\varepsilon}{D}\right] + 1.14$$

Gas pipelines always operate in the rough-pipe regime, in which Reynolds number ($\mathrm{Re}$) and kinematic viscosity ($\nu$) have no effect. This equation is close to the limiting form of the Colebrook-White equation at high Reynolds numbers; as with smooth-pipe flow, we recommend using one of the general equations, Colebrook-White or GERG, to avoid any numerical artefacts which might otherwise arise from switching equations.

## *Between C and D: smooth-pipe or rough-pipe turbulence?*

Around 20 000 < Re < 100 000 there is zone in which behavior is intermediate between *smooth-pipe turbulent* and *rough-pipe turbulent*. Both the relative roughness of the pipe wall ($\varepsilon/D$) and the Reynolds number of the fluid flow ($\mathrm{Re}$) have an effect. This region is glossed over in most introductory references, yet it is where most pipelines operate. What exactly is it that happens in the regime between smooth-pipe turbulence and rough-pipe turbulence? Colebrook and White's experiments indicated the presence of a gradual transition. It remains the most popular correlation among pipeline engineers:

$$\frac{1}{\sqrt{f}} = -2 \cdot \log_{10}\left[\frac{\varepsilon}{3.7 \cdot D} + \frac{2.51}{\mathrm{Re} \cdot \sqrt{f}}\right]$$

This correlation says that in that vicinity, roughness ought to have *some* effect as soon as we are in a turbulent regime, but perhaps not much effect. Two terms trade off. A *rough term* involves relative roughness, the ratio of roughness ($\varepsilon$) to inner diameter ($D$):

$$\frac{\varepsilon}{3.7 \cdot D}$$

And a *smooth term* involves Reynolds number ($\mathrm{Re}$) and the friction factor ($f$):

$$\frac{2.51}{\mathrm{Re} \cdot \sqrt{f}}$$

Those two individual terms are similar in form to Prandtl's smooth-pipe correlation and Nikuradse rough-pipe turbulence correlation, with slightly modified constants. The friction factor ($f$) is something like 0.01 or 0.02; so if Re is 40 000 the smooth term is:

$$\frac{2.51}{40\,000 \cdot \sqrt{0.01}} = \frac{2.51}{40\,000 \cdot 0.1} \approx 0.001$$

A pipe whose internal diameter is 1 meter must have a wall roughness of 500 microns to have an effect in the same ball-park as the smooth pipe term. This is much rougher than a real pipe wall, so we can say that at a Reynolds number of 40 000, the effect of wall roughness is negligible. Since this expression takes a logarithm, saying *"an effect in the same ball-park"* means *"doubling what is inside the log()"* to have a small but noticeable effect. When Reynolds number is around a million or more, even a realistic wall roughness that we expect in a typical commercial pipe will become consequential.

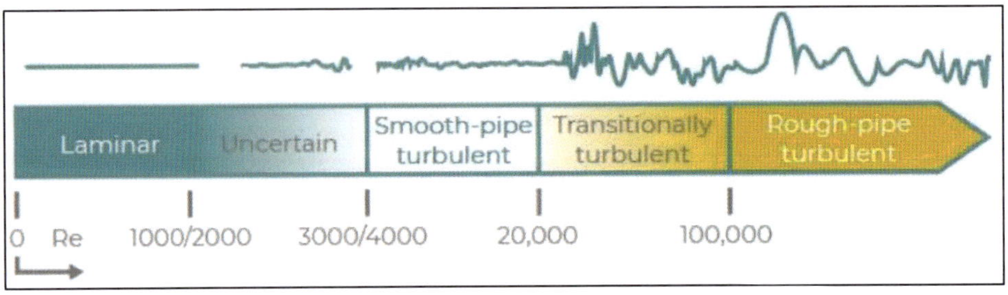


Figure 85 – How streamlines look in each flow regime, as the Reynolds number (Re) varies. The Colebrook-White correlation gives a smooth transition from smooth-pipe to rough-pipe turbulence, but in recent experiments it has been shown that this does not happen. Instead, what is observed is a sudden, sharp change between smooth-pipe and rough-pipe turbulence.

Unfortunately, recent experiments indicated this transition is not gradual but *sharp*, such that the relative roughness of a pipe wall kicks in *suddenly* at a middling-to-high Reynolds numbers (above 40 000). Nikuradse observed a sharp inflection point between the two regimes, contrary to others observing commercial steel pipes to show a smooth transition here. Now, experiments seem to be reaching his conclusion again.

As everyone isn't really sure when (or if) this transitional regime occurs, the European gas research group (GERG) came up with a novel way to deal with this controversy in the 2000's. They formulated an expression that lets the user make the transition as sharp as they want it to be via an adjustable parameter. They adapted the popular Colebrook-White correlation to equip it with a variable sharpness ($n$) that the user can fit to suit their needs. With $n = 1$, this produces a smooth transition almost exactly like the Colebrook-White equation; with $n \geq 10$, it produces a very sharp transition.

$$\frac{1}{\sqrt{f}} = -\frac{2}{n} \cdot \log_{10}\left[\left(\frac{\varepsilon}{3.7 \cdot D}\right)^{n} + \left(\frac{1.499}{\mathrm{Re} \cdot \sqrt{f}}\right)^{0.942 \cdot n}\right]$$

Why should we get to choose our own reality? Purveyors of fluids often find themselves compelled to put their hands up and seek a rule of thumb for their application to get it working! Any correlation describing the transitional regime is one unknown amongst many affecting the accuracy of a calculation. Some of these intractable questions are being addressed using deep learning approaches. Turbulence-scale Navier Stokes is the topic of some of this research, and it would be surprising if progress couldn't be made using the tools that have tackled more complex problems such as protein folding!

In a pipeline, we have what boils down to a one-dimensional flow, so our task is quite reasonable. Good engineering correlations have gotten us a long way, and pipeline simulators make verifiably sound predictions of how fluids flow. The Colebrook-White correlation for friction factor has proven to be quite adequate for our practical purposes.

# Flow of gases: steady states versus transients

All of the aforementioned calculations are premised on major assumptions. They often go unstated as *"it goes without saying"*. But for clarity, we pause here to say that our efforts so far have been striving to describe *steady states.* That is: flows and parameters unchanging over time. Pipelines, in reality, might be in a *transient* state, changing over time. In fact, gas pipelines never practically approach a steady state during the course of their operation; they are constantly subjected to imbalanced inflows and outflows, causing *linepack*.

The same friction factors are applied in transient as in steady states. Research on water hammer suggests that for liquids more complex expressions for friction might apply in the middle of a violent pressure surge, but we shall cross that bridge when we come to it. The analytical expressions for flow of gases through pipes are more involved and we won't cover them here. Suffice to say, the flow of a gas or indeed any compressible fluid is quite different to liquids.

In dedicated chapters of the main book, we look at shut-in pipelines, steadily flowing pipelines and transient operations in pipelines. But before rushing on to those technical discussions, let's take a closer look at the fluids carried along various pipelines.

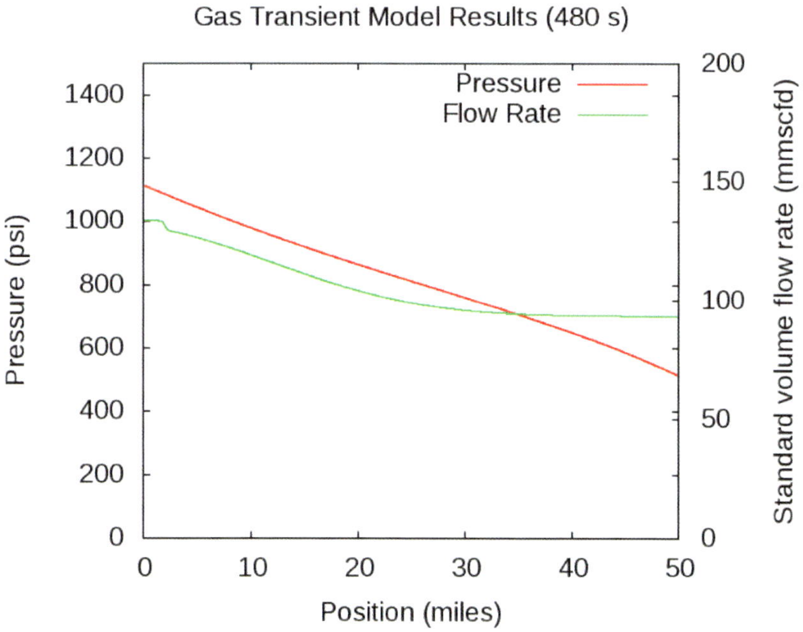


Figure 86 – A snapshot of a transient in a gas line; flow and pressure profiles evolve over time

# Appendix D – What do pipeliners do?

## D1 – Why we need pipelines

In this brave new world, why can't copper be our savior? A healthy skepticism towards the status quo is a hallmark of any healthy mind, and today it is more of a roaring tiger, presenting courageous challenges to everything we do. Who needs these old-fashioned technologies; *"energy efficiency is so 1970's"*!

Engines in a farmer's field can provide standby power if – and only if – the power grid and the gas grid can both handle it. An oilfield can only avoid flaring associated gas as *"stranded"* if – and only if – there is an economical route to market. This means a gathering network to a processing facility, followed by a pipeline (if within a few hundred miles) or another technology (further afield). A hydrogen economy of any color can come to fruition if – and only if – it can be transmitted, either as pure hydrogen or blended with natural gas. To manufacture hydrogen in an environmentally responsible manner, the carbon dioxide captured from its process can be transported to users or to burial if – and only if – we have pipelines.

The United Nations' most optimistic projections for a global energy revolution predicts a world ever more reliant on pipelines. Both electrical and pipeline infrastructure remains at the forefront of every energy transition we attempt.

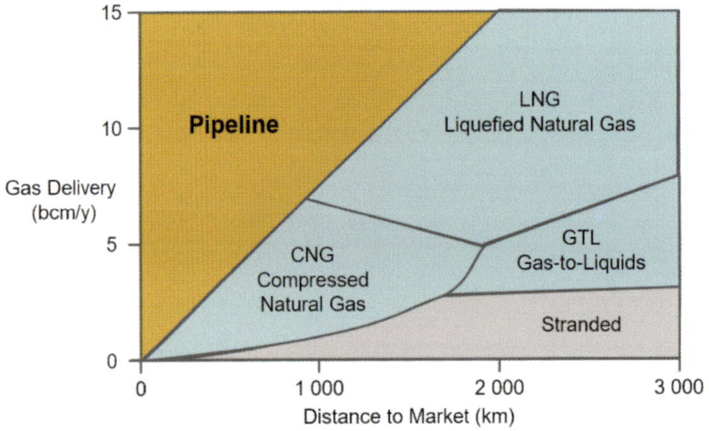


**Figure 87 – Economically preferred options for monetizing stranded natural gas depend on annual volume being delivered, and distance to consuming market (adapted from source).** [27]

At Atmos, our corner of this changing world is the pipeline itself. Whatever that pipeline might be transporting, we have assembled a team dedicated to modelling how fluids behave during pipeline operations. The pipeline is the role we play in serving and safeguarding people, industry and the environment.

Some say a power grid is the largest machine ever built in history, but pipeline networks definitely give those making that estimation a run for their money.

A network of gathering lines brings crude oil and associated gas from oil wells – or natural gas and associated liquids from gas wells – to central processing facilities where they are separated. Then, long-distance transmission pipelines (over three million miles worldwide) ship oil and gas at high pressure, with further midstream pipelines transporting liquid products from refineries to tank farms and low-pressure distribution networks. When we start looking at distribution networks, the numbers become truly astounding. The US alone has millions of miles of gas mains and low-pressure distribution pipelines. [49]

In this chapter we examine the fluids transported by pipeline, and why we use pipelines to transport them. In the times we live in, it is jarring for some people to meet someone in the *"tech"* sector working with infrastructure that is not virtual or electrified. To add insult to injury, the word *"pipeline"* seems to have been reappropriated by computer scientists for other purposes.

# D2 – BAD IDEAS AND LESSONS LEARNT

*"What of my Dross thou findest there be Bold*
*to throw away, but yet, preserve the Gold.*
*What if my gold be Wrapped up in Ore?*
*None throws away the Apple for the Core!"*

These lines quoted from a wonderful poem are how Trevor Kletz started his seminal work, beseeching those involved in the industries around him to heed his call. *"What Went Wrong?"* is a must-read for anyone who does anything near, with or about industrial equipment. By advocating for reason, he single-handedly popularized, or basically largely *invented*, the field of process safety.

*"Safety is yawn-inducing"*, we paraphrase him saying, *"because it has been reduced to hard-hats and days without incident. When actually, it's about designing and operating facilities not to explode"*.

Not to be disparaging towards occupational safety, hopefully you get the gist! Whenever we look back at the history of *any* industry, not just pipelining, we are faced with a poignant reminder of our past shortcomings as a profession. Engineering, all its disciplines, was born out of a series of significant mistakes. In the early days, there was a great deal ahead for the pioneers to learn. They were taught a host of valuable lessons the hard way. By trial and error, ambitious visionaries attempted approaches which turned out to be foolish. From their oversights, we learnt. Sharing these lessons isn't lip service, or a mere inconvenience. It's how we avoid falling for those same mistakes again!

Process safety talks a lot about fires, explosions and toxic dispersion because *those events happened*. They happened to people as well-meaning as you and me. In fact, they continue to happen, though society has grown curiously accustomed to them. Aside from specialist journals and local press, incidents go largely unreported. Seldom does a major incident have no predecessors. Near-misses happen all the time, but when is the last time we learnt from one?

We don't strive to do the impossible and eliminate risk entirely. What we <u>can</u> do is to take every sensible precaution, reducing risk to acceptable levels, which is defined in regulatory and ethical terms to be as low as reasonably practicable (ALARP). In other geographies, the threat of being sued alone offers a stronger imperative to avert harm. Although corporate culture hasn't evolved a sentient conscience to counteract the short-term profit motive, one shouldn't nonetheless discount the legal and financial imperatives which have proven enough to motivate against the worst transgressions in many contexts.

Progress can be measured in several ways. When something is shiny and new it's awfully tempting to push it to its very limit. But all that glitters is not gold. Today's expectations are (rightly) that we don't just do great things, but demonstrate that what we're doing is reasonable, safe and resilient. This involves forays into psychology and management, the realm of human factors.

Let's give a solid example. It is often through copious references to specific anecdotes that engineers think, and in this way Kletz made strides on the topic.

An operator of a crude pipeline might have picked up, through experience, that slack occurs during a shut-down. So when pumping commences at the start of their pipeline, they come to expect a long wait before flow appears at the other end. This wait becomes baked into their *de-facto* operating procedure, with an implicit *hope* that no leak happened during the period they spend waiting.

In 2010, a pipeline traversing Michigan was breached as it crossed a tributary of the Kalamazoo river. The National Transportation Safety Board (NTSB) investigation attributed the cause to a combination of factors, including corrosion fatigue and cathodic protection tape coming loose along a few feet of pipe. This raised alarms in the control room, but staff were unable to interpret them with certainty. For eighteen hours, they remained unaware of the breach. For much of that time, operators had shut down the pipeline, so the bitumen mixture escaped comparatively slowly. But as they attributed the alarms to a possible bubble in the line, they explained away anomalous pressure readings, and restarted the line twice. For a few hours in total, material escaped much faster. As a result, not only was this incident the first significant spill of diluted bitumen in the US, it also became the country's *largest* on-land spill to date. [115]

When plastic water pipes happen to run through oil-soaked ground, the water they carry may become contaminated with oil. So a spill of this nature and this magnitude was devastating, reportedly posing a direct risk to surrounding communities.

The purpose of presenting this anecdote is not to say that avoiding all incidents is our *only* lesson learnt, though there are interesting questions around how technical issues are communicated and translated into management decisions. It's to say pipeliners must be well-prepared with effective procedures and sound understanding to *mitigate* any incident if it occurs, however unlikely. An entire field of specialization is devoted to the psychology of a control room and the operators within it. Like an aircraft cockpit, many factors are at play. One principle is that if the decision maker is continually overwhelmed by false alarms or conflicting information, it inevitably affects their operating strategy.

As normally functioning non-psychopaths, we occasionally pause, take a good hard look at ourselves, and take pride in what we do. An attitude of continual improvement, always striving to do what we do better, is a force overcoming obstacles to progress. The *inertia* one encounters when doing this: the *"we have always done it this way, just because"* attitude, is an obstacle but not one that is insurmountable. Some teams do embrace continuous improvement.

Pipeline incidents, like any industrial accident, occur when there are series of failures in an *overall system*. These failures might be in some technical aspect of the pipeline itself, in its operation, or in the management processes meant to monitor and protect it. A good enterprise of pipeliners continually improve the technologies that inspect and monitor pipelines, using safety management systems to identify, prevent and manage gaps in their operations that can impact safety. They live by a strong safety culture in their day-to-day work. [28]

Many technical practices were born out of bad ideas we paid for in the past. Here are a couple of examples. *"It is now widely recognized that we cannot prevent fires and explosions by eliminating sources of ignition. We do what we can, but they are still liable to turn up. So we try to prevent the formation of explosive mixtures. Similarly, it is now argued that we cannot eliminate all causes of cracks, and so we should make sure that any that do occur do not propagate."* Major accidents are almost never caused by a mysterious technical mechanism that baffles engineers. Disasters happen due to workplace culture: ideas and habits that become deeply ingrained. People's bad mindsets can kill.

In 1974, the Flixborough disaster, caused by a hasty *"temporary modification"* to pipework on a process plant, ended in an explosion that killed 28 people. It again demonstrated a failure not only of technical competence, but of wider company processes. When the pressure rose a little above normal level, an expansion joint failed, but after this serious incident, nobody learnt anything new at all about expansion joints. The response from many companies was to introduce policies around *management of change* (MoC)! [29]

> *"Few engineers have the specialized knowledge to design highly stressed piping. In addition, the engineers at Flixborough did not know that design by experts was necessary. They did not know what they did not know."*

Two years later, a toxic chemical release took place at Seveso in Italy. These two incidents justified a systematic approach to process safety across the process industries. Governments regulated plants processing or storing large inventories of hazardous materials – COMAH, and the 1982 Seveso directive.

In 1988, an explosion and fire on an oil platform in the North Sea, Piper Alpha, cost the lives of 163 people. The immediate cause was identified: poor isolation during maintenance. The true cause was far more subtle: workplace practices. The official report concluded *"that the operating staff had no commitment to working to the written procedure, and that the procedure was knowingly and flagrantly disregarded".* [29] So the lesson learnt is far broader than the mere existence of a permit-to-work system. The lesson learnt concerns psychology and management: how do we ensure every member of our operating staff is committed to working to written procedure, that they hold it in high regard?

We have much to learn from other sectors. Manufacturing adopted principles that seem simple but have had a great impact: *Kaizen* and *Lean – Six Sigma*.

- *Kaizen*: continuous improvement, where every team member strives to improve how work is done as a matter of normal everyday business. Individuals always seek efficiency, instead of waiting to react to events
- *Lean – Six Sigma*: problem solving techniques with a culture of waste reduction, improving operational performance by avoiding variance, achieving fewer defects via efficiently controlled working processes

In software one finds themselves surrounded by an ocean of *"Agile"* buzzwords that in essence strive towards the same objective: adaptable ways of working. Having the humility and self-awareness to critically examine what we're doing.

Central to any effective workflow is an acknowledgement that *we might be wrong*. This isn't a personal dig against the integrity or worth of another person, to be received with sensitivity and indignation. It is an inevitable part of any process of creative innovation. What matters is that we make the appropriate caveats, and have procedures in place for quality assurance that ensure our engineered solution is fit for purpose. And, of course, that we seek feedback, act on it in a timely fashion, learn lessons, and improve with every iteration.

The process industries in general, and the energy industry in particular, rightly or wrongly earned themselves a bad reputation in the popular mindset. A macho culture resistant to change is slowly learning the hard way to strive to do better every day rather than coasting along indifferently *"business as usual"*.

# D3 – GOOD IDEAS AND THE STATE OF THE ART

Pipelining projects are delivered according to rigorous processes. Overarching principles include inherent safety, seeking and adopting less risky ways to reach the same objectives. To keep their pipelines running safely, reliably and economically, pipeline enterprises employ professionals and call on specialist consultants. A variety of independent experts all work in complementary roles.

*Dispatchers*, operators or controllers * monitor the pipeline's daily operations, deciding what to do next, and intervening to override the automated systems where needed.

* Which term is preferred and what each denotes varies by company and country. *"Operator"* is an ambiguous term: it can refer to a whole pipelining enterprise. And there is ambiguity in *"controller"* too, as that can refer to control systems.

They choose the setpoints for units, perhaps every hour and every day. In ordinary circumstances, any actions more frequent than that are automated.

*Operations planning* experts support several stakeholders, including the dispatchers, and commercial teams. They determine whether it's possible to meet some proposed commitment. Nominations (supply or demand flows) on a gas pipeline or capacity on a liquid pipeline is agreed every month. This team is also responsible for putting procedures in place for the dispatchers to follow.

*Business development* teams negotiate the line between commercial contracts with suppliers and customers, reconciling profitability with the physical parameters of their pipelines' performance. They make big, long-term strategic capital expenditure decisions that happen on a time-scale of years or decades.

*Consultants* provide outside expertise, and may be employed for detailed tasks like conducting *"flow assurance"*[*] [30] or surge studies. *Designers* specify how a pipeline is to be constructed, commissioned and operated. They examine, in detail, the pipeline's characteristics and its expected performance in various operating conditions. They may be consulted on *optioneering*, deciding a best course of action when weighing up several proposed ways to tackle a problem.

A pipeline simulator suite supports many of these tasks, enabling teams to effectively design, build and operate pipelines that continually drive progress in the field. That software is itself delivered according to rigorous processes, and audited for the duties it is responsible for.

Within a pipeline simulation vendor, a dedicated team including developers, managers, engineers and all manner of technical experts work to improve the performance of the pipeline simulation software, drawing on experience and collaborating with their users. The innovations that are happening are not just technical calculations.

Meeting the needs of all those different kinds of users has seen a step change across this industry. Ultimately, pipeline simulation is much like any industry in the 21st century. Raw data is available, but overwhelming. We add value by processing it, analyzing it by applying our technical expertise using the right technologies, and presenting the relevant results in an appropriate manner to every stakeholder. A pipeline simulator is useful if it fulfils that end-to-end role.

---

[*] Flow assurance is a loosely defined term that may encompass all issues important to maintaining the fluid flow from a pipeline's inlet to its outlet. It is a widely used term, especially for hydrocarbon pipelines, and it is possible to define the system boundaries (*"inlet"* and *"outlet"*) to include or exclude upstream production facilities or reservoirs, and downstream processing or distribution.

# D4 – KINDS OF PIPELINE

It is exceedingly rare that a pipeline is completely repurposed from carrying one fluid to conveying another. In the industry's parlance, a line is described by its *service*; the fluid it transports, and often by its position in the value chain: gathering lines from a well-head (for hydrocarbons), high-pressure lines called transmission lines, mainlines or trunk lines, and low-pressure distribution lines. The following subsections, not attempting to be exhaustive, will run through most of the widely transported pipeline fluids driving entire economic sectors.

Gases and liquids behave differently, and their driving economics are distinct. It is helpful to regard gas pipelines as a category separate from liquid pipelines. One should bear in mind, though, that while gas distribution pipelines are tested with compressed air before entering into service, high-pressure gas transmission pipelines are filled with liquid water for hydrostatic testing. Conversely, a liquid-service pipeline is purged with nitrogen gas. Moreover, during operation, gas pockets might form in a liquid line, or liquid might drop out in a gas pipeline, as we see in a dedicated chapter of the main book: *"Beyond single-phase"*.

Although repurposing a pipeline from liquid service to gas service is rare, it can and does happen. A good simulator shares the same underlying data-model for a liquid and a gas pipeline, restricting liquid and gas model items for the user's convenience only. This compatibility saves rework if repurposing were to be investigated. Who knows during our lifetime how many liquid lines go to gas service! More to the point, a good physicist, a good programmer or a good person all round ought to share an algorithm if it applies equally to liquids and gases alike. There's no sense in making something a special case when it isn't.

The US has pipelines twice as long as it has paved roads: 216 000 miles for hazardous liquids, 319 000 miles for gas transmission, 400 000 miles for gas gathering, 2 million miles for gas distribution and a million miles of water mains.

# D5 – PIPELINE LIQUIDS

In *almost incompressible liquids*, pressures transmit rapidly, and even the smallest changes in elevation are influential. Three kinds of pipeline broadly fall into this category: water pipelines, crude pipelines and product pipelines.

## Water pipelines

**Figure 88 – A water pipeline supplying a major city in Australia**

Water is famously almost constant in its density at one ton per cubic meter, or a kilogram per liter. This is true at most temperatures, except near freezing, where anomalously water's density *falls* slightly as it cools. It is the only fluid to exhibit this behavior, which happens for the same reason that ice floats.

In developed countries, water pipes supply almost everyone, though standards remain lax in places. Only the remotest reaches still rely on boreholes, tankers and jerry cans. But the water sector remains curiously indifferent to rampant leaks. These grow to causing huge sinkholes surprisingly often, even in cities!

Essential to life, food, and countless industries, there is a slow but sure shift in how we deal with water. No longer *"just a utility"* or a casually chosen process medium, we no longer take its availability and quality for granted. Entire countries and global cities now rely on desalination and water recycling. It has become a strategic resource which warrants care and attention in sourcing, treatment, usage and transportation alike.

Where water scarcity has become a more obvious reality, particularly in arid climates, operators take great care when pipelining this precious resource. Desalinated water, for example, traverses long distances to reach people reliant on its supply, literally as a matter of life-and-death. Several industries are water-intensive, including mining and oil & gas extraction, for which reducing water consumption is a major priority. Pipelines supply some of these facilities with water, often saline groundwater or brine, over long distances. Pipelines also transport their effluent *produced water* to treatment plants.

Water distribution networks operate at a *steady state*. They don't need a transient simulator – so long as their woefully common leaks are happening at a constant rate, it never hindered their modelling. That industry finds the steady approach to be fine. Water pipelines pose unique challenges that make them more complicated to operate than one might expect. A section that sits out of use for too many days becomes unpalatable due to bacterial growth: illegal to deliver as potable water. Eventually, it accumulates impurities that are liable to forming methane and hydrogen, posing an explosion risk! So operators are naturally interested in tracking the *age* of water in their system. Their routine work includes creating *flushing plans* for their water networks.

Another challenge is that these extensive networks often lack instrumentation. In places, they seem to even lack shut-off valves, and a water leak in Manchester might completely drain a reservoir before someone figures out how to stop it. As metering improves, we are presented with an opportunity for smarter modelling. A complete digital twin of trunk mains and distribution networks helps operate efficiently and safely. [37] In an era of water scarcity threatening even capital cities in rainy climates (such as London) with drought, incessant leaks are becoming intolerable. Ageing infrastructure is being re-examined in an effort to conserve this vital resource.

# Why are liquid hydrocarbons pipelined?

Even in the United Nations' most optimistic projections for a transition away from fossil fuels, there remains a need for expanded pipeline capacity. If one recognizes there is a supply and demand for these hazardous substances, the task becomes providing the safest way of transporting them. If that recognition is a point of contention, we go back and talk about the liquids, not the pipeline. Essentially, if one is comfortable with society using refined liquid products in the first place, the ethical question of pipelining them then seems uncontroversial. Rather the opposition to pipelines might arise because:

- Neighbors fear a spill will affect them or their local environment
- Due process was violated so the opposition is a matter of principle
- Or opposing a pipeline is really about opposing liquid fuels themselves

Although a pipeline makes transporting these fuels cheaper, the demand doesn't vanish when no pipeline exists: the consumer is just exposed to more risk, and also pays more for it. A counter argument to good infrastructure is that it *"creates induced demand"*, but we see the world over that people need energy. Where pipelines are unavailable, shippers are paid to resort to other means of transport – rail and road – posing an immediate risk more severe by an order of magnitude. The devastation wrought by pipelining disasters is readily attributable to poor practices and lax regulations, not pipelining itself.

Pipelining liquids is forty times safer than transporting them by rail, and one hundred times safer than using road tankers. Unlike other infrastructure, a pipeline isn't shared by the general public: a good right-of-way is chosen to be kept as far away from people as reasonably possible. If the world's major oil producers did not have pipelines, this would introduce thousands of extra tank wagons onto rails or roads near people. Every single one would be considered a pressure vessel in its own right, with individually dedicated equipment. [39]

**Figure 89 – Derailed tanker train (National Transportation Safety Board)**

To put these statistics into focus, major accidents have occurred where a lack of pipelines or pipelining capacity has been forcing shipments of hydrocarbon liquids to be delivered by rail. In 2013, a freight train carrying crude oil derailed in a populated region of Quebec, causing a tragic loss of human life. In 2020, a freight train carrying diesel fuel derailed in Wales, prompting evacuation of nearby homes and devastating the environment. Such derailments seem to keep happening every few months, but near-misses are peculiarly devoid of public attention. And roads are, of course, no safer than rail, quite the contrary.

# What does a fluid do? The parametric approach

We know what water is composed of: $H_2O$. But a batch of hydrocarbon liquid, be that a crude oil or a liquid product, is typically poorly characterized. Liquids cannot be directly measured for composition like gases can. Their component breakdown is unknown, or at best, difficult to determine in real-time. When we don't know what a fluid is precisely made of at any given instance, it is dealt with via a bulk approach, and we refer to it in Atmos SIM as a *parametric* fluid. Substantial variation in a fluid's composition can be characterized only by bulk parameters, like its density and its viscosity at a set of reference temperatures.

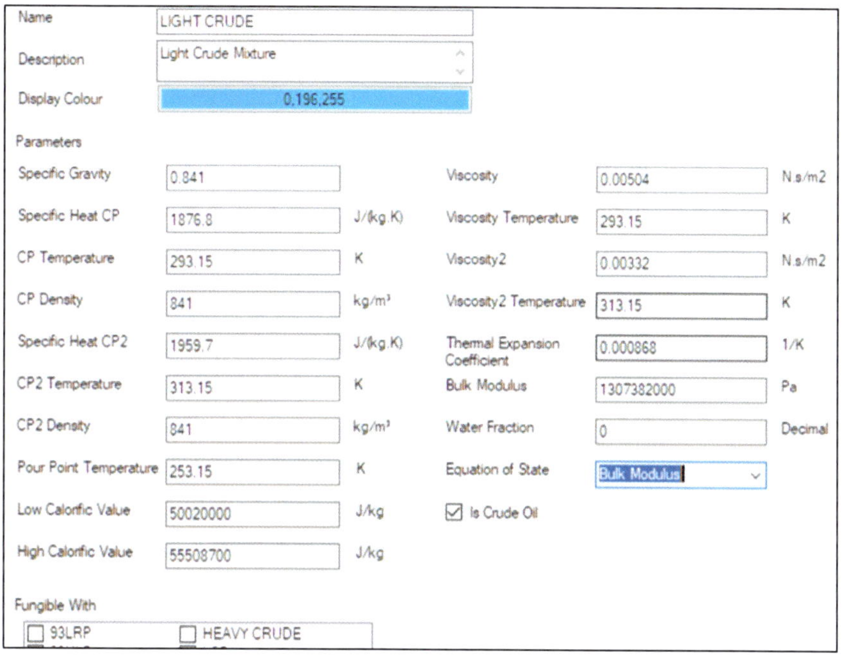


**Figure 90 – An Atmos SIM dialogue defining the properties of a parametric liquid**

Accuracy aside, the parametric approach might appear to be a simpler method. But it presents its own complications, particularly when considering mixtures. At a mixing point where an incoming branch joins a mainline, what do we do? To handle that situation, we closely consider matters like viscosity blending. Taking a weighted average turns out to be a poor approximation. If we aren't careful we draw the wrong conclusions about how a fluid behaves in a pipeline.

# Crude oil pipelines

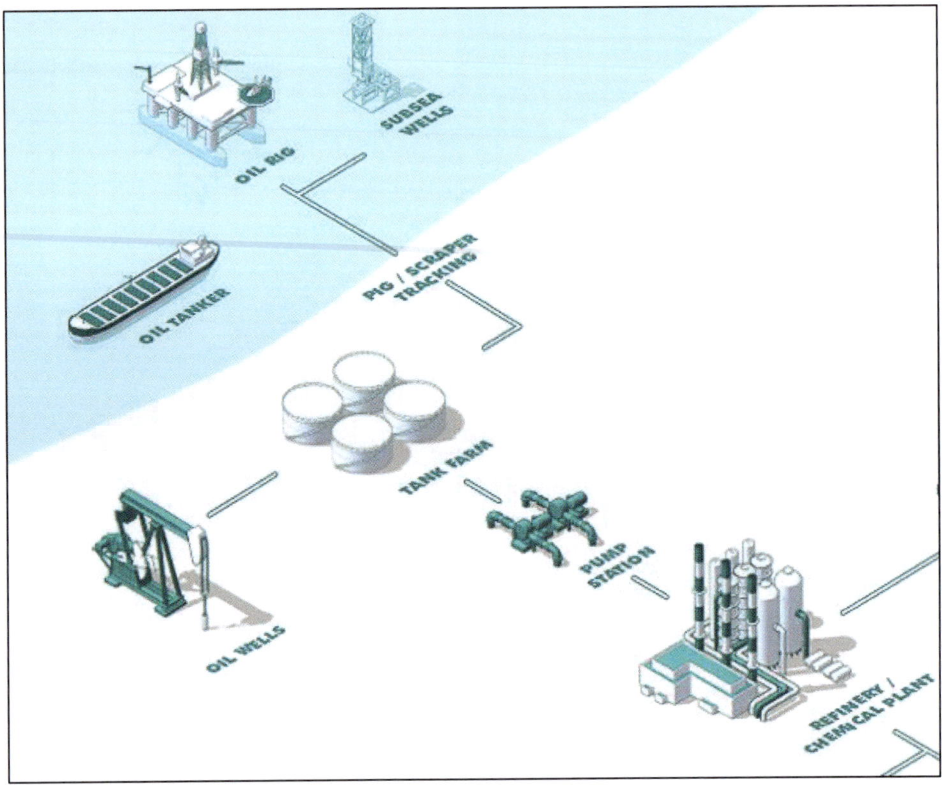


Figure 91 – A schematic of how crude oil pipelines link wells, terminals and refineries

Crude oil comes in many grades, varying from thin, brownish or greenish hues, to thick, black oils resembling melted tar. It is characterized by a standard set of parameters, varying well-to-well and by origin, described by vague terms like *waxy* and *sweet*. Its parameters are often uncertain: the operator might not know the properties of a batch they're shipping to any reasonable accuracy.

Would you be able to pick up a barrel of crude oil?

An oil barrel is 158 liters in volume, which is 0.16 cubic meters. A conservative density of 700 kg/m$^3$ puts its weight at over 110 kg, and a denser 950 kg/m$^3$ has it pushing 150 kg. Even if you lift with your legs, this seems an unwieldy object to place onto a trailer, though I won't say you can't do it if you watch your back! Indeed, people used to do exactly that. The first modern crude pipelines were an epic tale, with rival mobs of teamsters canoeing barrels down river rapids.

A given grade of crude oil has a density that varies with temperature as well as operating pressure, so engineers working with a crude or refined product refer to its *relative standard density*, which some call specific gravity (SG). In the case of liquids, it is defined as the ratio of the fluid's *standard density* ($\rho_{\text{std}}$) to that of water. Specific gravity is dimensionless if all units are consistent. It is used as an (imperfect) surrogate for a crude oil's quality, along with basic sediment and water content (BS&W).

Pipeliners of hydrocarbon liquids then do something yet more complicated, and use an American Petroleum Institute quantity called *"API Gravity"*. This is an equivalent but, confusingly, *inverse* measure. It broadly categorizes a given crude oil based on its density relative to water. It is defined as:

$$\text{API Gravity} = 141.5 \, / \, \text{Standard Gravity} - 131.5$$

Light oil is said to be *"high gravity"*, with an API Gravity of more than 10, floating on water as an oil slick, with some chance of being cleaned up, albeit at great expense. Extra-heavy oil is *"low gravity"*, with an API gravity less than 10, sinking in water, and so particularly devastating if released. To help rationalize this nomenclature, we could think of the oil sitting high or low in the water table.

**Figure 92 – Heavy, *low-gravity* oils are harder to clean up than light *high-gravity* oils (NTSB)**

The heaviest fraction of all is bunker fuel, *"the bottom of the barrel"*, laden with poisons and sadly today's prime choice for ships carrying the bulk of global trade. It is a *solid* at room temperature; its *"pour point"* at atmospheric pressure is warmer than the ambient temperature, so it must be heated before it flows.

# Refined products pipelines

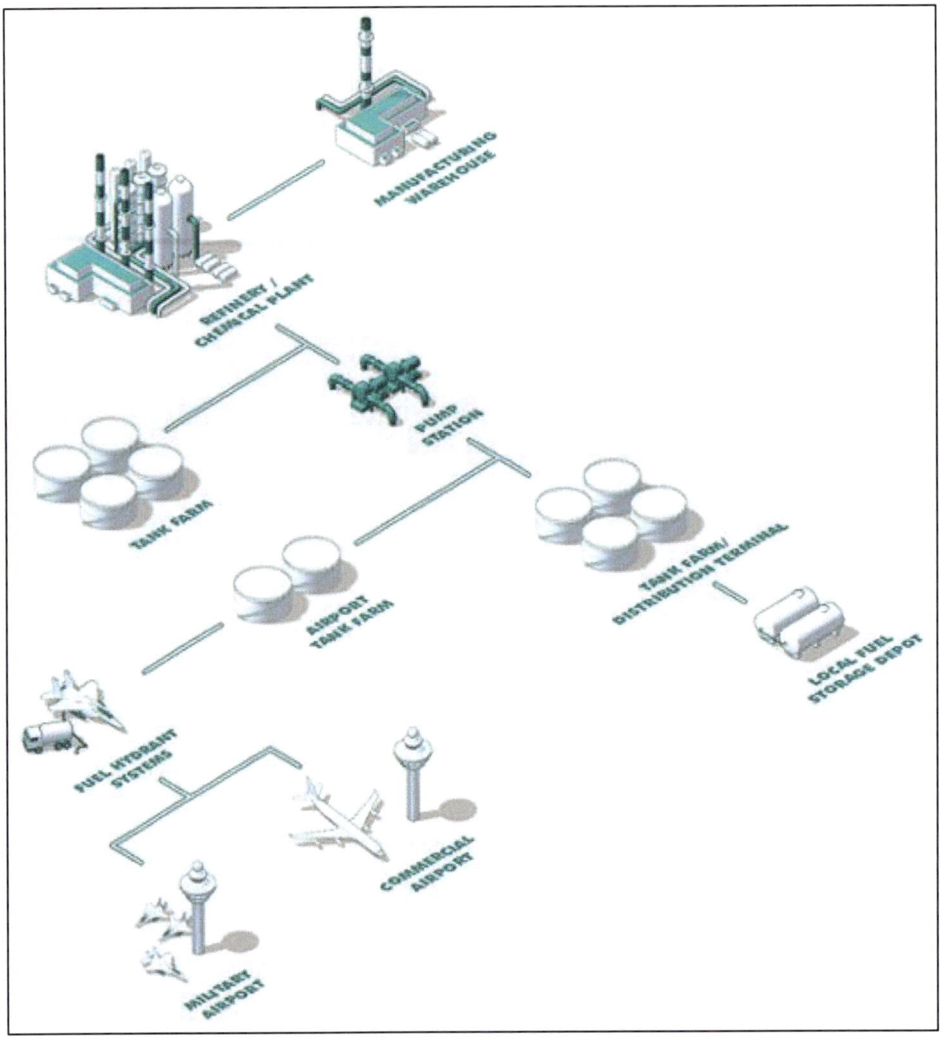


**Figure 93 – A schematic of how refined product pipelines link refineries to factories, airports and tank farms for onward distribution to local fuel depots**

Refineries convert crude oil into chemical feedstocks and liquid products. These are also made *"synthetically"* in gas-to-liquid and coal-to-liquid plants. These are used as fuels, and are blended and quality controlled to comply with local standards, such as Euro-5. Pipeliners refer to them simply as *"products"*.

Product pipelines can deliver these liquids over hundreds of kilometers. This kind of pipeline may transport a single product, such as jet fuel from a refinery to an airport, or it might be a multiproduct line batching different blends of refined products to a tank farm for onward distribution. These liquid product batches remain in the line for a substantial period along their journey, during which they might intermingle and form what is known as an interface.

To avoid intermingling, batches might be separated by solid devices called pigs that are inserted into the line, and travel along with the flow. Pigs used for this purpose are known as spheres.

Operations performed on product pipelines often revolve around:

- batch tracking (predicting arrival time, reporting custody, etc.)
- pig tracking (predicting arrival time, preparing valves, etc.)
- monitoring conditions (avoiding column separation, etc.)
- optimization (reducing power cost, increasing flow capacity, etc.)

Every liquid product is characterized differently according to how it is used, and many are even described differently country to country. A complication in some *multi-product pipelines* (MPP's) is that only some of the product baches are incompressible: petrol (=gasoline), diesel, jet fuel (=kerosene) and bunker fuel. While other batches in these same lines are compressible fluids, such as liquefied petroleum gas (LPG) or propane.

Many diesel engines are capable of running on their originally intended fuel, vegetable oil, provided they cope with its worse viscosity and calorific value. We have not come across pipelining used vegetable oil, but pipelines have moved batches of *biodiesel*. Associated compatibility issues were observed as jet fuel became cross-contaminated with fatty acid methyl ester (FAME). That made pipeliners careful with which batches are allowed to share the same pipeline.

Sometimes, a batch of fuel might be a unique blend. As well as satisfying the quality standards of the customer, and tracing every batch back to its supplier, the pipeliner bears in mind a further consideration. How can we tell where one batch ends and the next begins? The line between one substance and another is rather arbitrary, and by and large, a batch cannot be distinguished by density alone, as the adjacent batch might have an overlapping density range.

| Fluid – terms vary by operator | State | Density $kg/m^3$ |
|---|---|---|
| Liquified petroleum gas (LPG) | Liquefied | 520 – 550 |
| Natural gas liquids (NGL) or Y-grade | Volatile Liquid | 500 – 700 |
| Isomerate | Liquid | 640 – 720 |
| Light gasoline / petrol | Liquid | 710 – 730 |
| Middle gasoline / petrol | Liquid | 720 – 770 |
| A particular *"high-gravity"* light crude oil | Liquid | 700 – 750 |
| A different crude oil | Liquid | 700 – 800 |
| Kerosene or jet fuel | Liquid | 775 – 840 |
| Petro-diesel | Liquid | ~ 850 |
| Bio-diesel | Liquid | ~ 880 |
| A viscous heavy crude or diluted bitumen | Liquid | < 940 |
| Water | Liquid | 1 000 |
| Bitumen | Liquid? | 1 010 – 1 060 |
| Milk | Colloid | 1 030 |
| Mining Liquor | Sludge | 1 250 |

**Table 5 – Typical density ranges of various liquids at standard conditions (standard densities)**

Automotive fuels are categorized as petrol (gasoline) or diesel. These are each subject to their own grades. Petrol is broadly described by an *octane number*, whose definition differs by country: 95 or 98 means something different in Europe than it does in America. Diesel is described by a *cetane number*. It was promoted as cleaner-burning in many countries as it emits less carbon dioxide ($CO_2$), though singling that out and ignoring other harmful pollutants was regrettable in hindsight. The *diesel-gate* scandal rocked that claim when it emerged diesel engines often exceed stated limits on nitrogen oxides ($NO_x$). Trace components matter in exhaust emissions, and in pipelining as well.

# Refining liquid fuels consumes hydrogen

The precise specifications defining the quality of a liquid fuel are becoming more stringent to both improve air quality and protect the environment. For example, petrol (gasoline) now contains less aromatics and olefins than it did a decade ago, and diesel in places must contain less than 10–50 ppm sulfur.

> *"These trends result in an increasing hydrogen content in the fuels (the H:C ratio is approaching 2). This has been achieved simultaneous with the incoming crude slate becoming heavier, with more sulfur and metals in it. Those trends together require more hydrotreating and hydrocracking to happen in the refining processes."* [31]

Refineries can supply some of their own hydrogen. It's produced incidentally by catalytic reformers and in certain off-gases. A refinery's hydrogen balance depends on the slate of incoming crude, and the specification of its products. Some produce a surplus. Others are in deficit and so make or import hydrogen.

Figure 94 – Pipelines at an oil refinery in Antwerp, the Netherlands

# Tracing contaminants, origin, and more!

We saw that we can characterize a pipeline fluid by one of two approaches: the *parametric* approach, or the *constituent* approach. In both approaches, it can be immensely helpful if the user can also define *trace* components. Those have no bearing on pipeline hydraulics, but prove useful in several other ways.

One important way the trace components feature is exploited is for composition tracking of contaminants, non-bulk materials present only in minute amounts. Keeping track of those helps operators act to maintain them at acceptable levels by blending or dilution. Trace amounts of water, for instance, might be present in many oils.

One dangerous contaminant is hydrogen sulfide ($H_2S$). It is measured in parts-per-million (ppm). Upstream oil and gas producers make a distinction between *sour* oil and gas, which contains hydrogen sulfide (>16 ppm), and *sweet* oil and gas, which doesn't. A field producing sweet oil and gas can, over time, gradually become sour, its fluids requiring more sweetening. This must be detected in time, as it poses a serious risk to life and unexpected corrosion.

Trace components also prove useful for origin identification, a particularly handy capability when tracking liquid batches that happen to be *fungible*.* Sometimes two liquids seem identically equivalent, but a contract specifies that the molecules delivered must be the same molecules that they put into the line. This is done by making up a trace component for each possible origin, tracked with every batch, and initially assigning a batch to 100% from its own origin and 0% from all the other origins. As the batch moves through the system and mixes, it maintains the fraction from each origin until it is eventually delivered.

Trace components can be used for origin identification.

A trace component could be applied to a batch if good-quality data tells us that it is not up to specification, so operators know when it's arriving and prepare to divert it to a separate tank. Or another subtly different situation can arise.

Trace components can be leveraged by pipeliners who know that a period of bad quality data happened at the inlet. The bad quality could trigger a signal, so that fact is tracked as a property of the batch that had been injected during that period. We wish

* Fungibility is how pipeliners say two liquids are mutually interchangeable.

to track a batch originating during bad quality data if, for instance, there's no density reading so we are unsure what fluid it actually is.

Trace components can be updated in the tracked fluid when it mixes with other fluid by any of several choices of *mixing rules*. One would wish to be extra careful with contaminants, and thus might choose to take the *sum* of their concentrations together if they mix at a point. For other purposes, a weighting based on flow rate may be suitable, such as for a trace component keeping track of calorific value that has been measured along the pipeline.

> Depending on a trace component's purpose, we select the mixing rule as a sum, highest or lowest value, or average.

# Diluent, bitumen and diluted bitumen

Tar sands (known as bitumen, in contrast to conventional crude oil) is a thick, viscous, highly corrosive liquid. Transporting it is facilitated by blending it with diluent, to make *diluted bitumen* (abbreviated to *dilbit*). The diluent is typically either light crude oil or volatile natural gas condensate. It has been pipelined regularly in Alberta, Canada since the 1980's, and has gained unprecedented attention in the 2000's as a lot of it started to be pipelined in the United States.

The mixture of these two fluids is challenging on several counts, one of which being that it is non-Newtonian, exhibiting shear-thinning behavior. Some dilbit pipelines carry unmixed bitumen and water along with the mixture, either down the core of the water or along the bottom of the pipeline, as a two-phase flow.

The US environmental protection agency (EPA) recommends pipelines that carry bitumen should no longer be treated like those carrying other oil. Existing regulations were never developed or reviewed with its properties in mind. It is highly corrosive especially at the high temperatures and pressures required (70°C, 150 atm). This makes it more likely to cause a leak, historically, and in that case, it is harder to respond to and contain than conventional oil spills. [32]

# Compressible liquids: liquefied flammable gases

Some pipelines carry highly compressible liquids, either as the only fluid, or as a batch amongst other liquid products. Compressible liquids are mixtures of light hydrocarbons which are volatile vapors at ambient conditions, but at typical pipeline operating conditions have liquid-like densities. These densities vary noticeably, so liquids such as propane or ethylene are said to be *"spongy"*. In 2015 the US was estimated to have 65 000 miles of pipelines carrying highly volatile liquids (HVL). [33]

Natural Gas Liquids (NGL's) are hydrocarbons removed at processing facilities from upstream natural gas wells. NGL's may vaporize, and broadly consist of ethane ($C_2$), propane ($C_3$), butane ($C_4$) and heavier fluids known collectively as pentane-plus ($C_{5+}$). Propane onwards consists of several isomers which affect the properties of a particular batch. Some call propane and butane mixtures *liquefied petroleum gas* (LPG), even if it originates from natural gas rather than petroleum. Others consider the term antiquated and prefer to avoid using it. These *"liquefied flammable gases"* are valuable. Propane, for example, is useful as a petrochemical feedstock for propylene production. It can be combusted in gas engines, but its energy value is low, so its value is higher than its combustion value. But these fluids are also challenging to pipeline. Propane is only a liquid at ambient temperatures if pressurized by 20 atmospheres or more. Such volatility makes a pipeline susceptible to going slack in low-pressure regions, as we shall discuss in a dedicated chapter of the main book (*"Beyond single phase"*).

An LPG pipeline caused a major disaster at San Juanico when it exceeded its maximum allowable operating pressure (MAOP). The over-pressure event failed to trigger a safety-critical pressure relief system. A vapor cloud caused boiling liquid expanding vapor explosions (BLEVE's), destroying the tank farm, killing 542 people. Indeed, a recent review by the UK Health and Safety Executive (HSE) compiled a list of vapor cloud explosion incidents around the world, in which propane pipeline incidents feature disturbingly prominently. [34]

Pentane-plus ($C_{5+}$, mainly $C_5$-$C_{10}$) includes *natural gas condensate* (NGC), and some call it *condensate*. Condensate pipelines must be pigged regularly, as these heavier compounds might form deposits. Some refer to condensate as *naphtha*. * US refiners reserve that term for the distilled product, whereas Middle Eastern and European pentane-plus is often referred to as a specific highly paraffinic naphtha equivalent, stripped directly from a gas stream to exclude ethane, propane and butane, to suit an

---

* Which derives its etymology from the historic term *naft*.

oil refinery's ethylene cracker. These liquids exhibit significant compressibility, their density affected by pressure, so they cannot be described as having a simple constant bulk modulus and thermal expansion coefficient. This means they require advanced, dedicated equations of state. Liquefied petroleum gas (LPG) is either stored at ambient temperature in high pressure conditions, or at refrigerated temperatures. The latter approach is often preferred for large storage facilities, because at low pressure there is a slower leak rate through a hole of a given size, and at cool temperatures a lower evaporation rate too. However, this latter approach introduces refrigeration and vaporization equipment, which presents their own new risks.

# Ethylene pipelines

Ethylene is an unusual pipeline fluid in that its critical point (50 bar, 9°C) lies well within a typical pipeline's operating conditions. This means it is often pipelined as a supercritical fluid: compressible like a gas, but with flow behavior akin to liquids. In the US, ethylene pipelines are regulated as carrying volatile liquids, and so are considered hazardous liquid pipelines. The most general equations of state are least accurate in the vicinity of any fluid's critical point. For this reason, ethylene has its own dedicated equation of state, and Atmos SIM can handle multiple equations of state in the same model to facilitate using it in a pipeline system.

Kletz in his *"Lessons from Disaster"* brings forward an example of a 12-inch ethylene pipeline. Ethylene flowed through it at a pressure of 90 atmospheres, and a double valve arrangement was closed to repair an instrument. When the nearest motor-operated valve (MOV) was reopened in a lapse of about 3 seconds, the atmospheric air pocket present between the valves was compressed by flowing ethylene acting as a piston. On adiabatic compression, a change in pressure ($P$) causes change in temperature ($T$) by this relation, where the exponent involves heat capacity ratio ($\gamma$):

$$T_2 = T_1 \cdot \left(\frac{P_2}{P_1}\right)^{\frac{\gamma-1}{\gamma}}$$

This translated into flowing ethylene heating air up to a temperature of 800°C. At this temperature ethylene started *decomposing* – a chemical reaction that is spontaneous at high enough temperature and pressure. It is strongly exothermic, meaning it releases more heat, which heats the ethylene further, causing further decomposition. The zone of decomposition travelled 80 meters upstream, at which point the velocity of the decomposing front equaled that of the ethylene flowing in the other direction. As a result, all of the decomposition happened at that one point along the pipeline, causing it to overheat and burst. [35]

# Carbon dioxide pipelines

There are 6 300 km of existing $CO_2$ pipelines in the US. Much has been learnt from decades of operational experience. [36] Carbon dioxide has at least three big users. They don't constitute a large enough market to use all our emissions:

- enhanced oil recovery (EOR), injected into wells to assist oil extraction
- feeding vegetables (tomatoes, peppers, etc.) grown in greenhouses
- manufacturing beverages (fizzy drinks), after obsessive purification

As carbon dioxide will never be disposed of by the free market, it is only usually recovered if regulations encourage it. Just as, say, emitting carbon dioxide in a steelworks is paid for by *carbon credits* funding carbon-offset projects planting trees elsewhere, those carbon credits could, via legislation, be traded and used for carbon capture and storage (CCS). Detractors argue that reviving government subsidies and investment into carbon capture is merely a way of further prolonging or at least *justifying* our continued reliance on fossil fuels.

Carbon capture and storage has previously been a temperamental technology in practice, and though proven to work at pilot scale, it was often prohibitively expensive to commercialize. So far, it has not yet been a viable way to bury carbon dioxide underground in any substantial quantity. Nonetheless, a lot of dedicated people are working on it and hoping it will help, particularly for those industrial producers of carbon dioxide whose chemical processes will never otherwise decarbonize. Projects are underway to store carbon dioxide in readily accessible rock formations: disused oil and gas fields.

When pipelining carbon dioxide, it is handled near or above its critical pressure of 74 bar, but below its critical temperature of 31°C. At these conditions, it is not supercritical, but is said to be in its *dense phase*. They are regulated in the US as hazardous liquid pipelines. Any sudden fall in pressure presents a significant hazard. [37] Accurate simulation of carbon dioxide with significant contaminants is difficult, because although there are good equations of state describing the density of pure $CO_2$, they are not so accurate for a contaminated mixture.

# D6 – PIPELINE GASES

## Why is natural gas pipelined?

In 2019, several natural-gas utilities in New York City stopped accepting new customers, citing jammed supply lines on the coldest winter days. Despite the prolific *Marcellus Shale* gas field lying a mere three hours away, the pipeline infrastructure was unable to cope. [38] Advocates of natural gas aren't naysayers insisting on fossil fuels to conspire against a eutopia of electric bliss. They're presenting a sensible transitions away from frankly ridiculous situations:

- Burning coal causing a permanent smog over some of our major cities, because plentiful natural gas cannot reach them to generate power
- Every building keeping a tank of diesel feeding its own off-grid gensets, because natural gas supplies and the electrical grid are not reliable
- Power grids even in advanced economies resorting to curtailment, because the power plants are not as flexible as gas-fired generators

For an extensive network, such as the one covering much of Northern Europe and the North Sea, the pipeline itself acts as a means of providing a *reserve store*, meeting customers seasonal, weekly and daily fluctuations. Just as a more extensive electrical grid is more resilient, an extensive integrated gas network is inherently more resilient.

Freeing natural gas from the pipeline by shrinking its volume 600 times allowed it to become a globally tradeable commodity against a global benchmark, Henry Hub in the US (and similar hubs elsewhere). But this comes at quite a cost! Without digging into the carbon intensity of liquefied natural gas (LNG) versus pipelines, one can safely shudder at the thought of just the sheer *power consumption* by refrigeration to -160 °C on such a mammoth scale. For deep ocean crossings, there's no alternative. But across land and sea, pipelines are a remarkably efficient mode of gas transport. Gas pipelines just make sense.

To cover distances within 3 000 km, it is more cost-effective and efficient to pipeline large volumes of natural gas than it would be to ship it liquefied. For larger-diameter lines the economical range can extend as far as 5 000 km. Around 2-3% of transmitted gas is consumed by compressors over the length of a typical pipeline, which is substantially lower than stated electric power transmission losses (6-7%). [39] Using larger transmission lines and clever neighbor-trading arrangements improves pipeline and electrical losses alike.

Gas distribution networks have become decentralized. Gas-fired generation now routinely happens at combined heat and power plants, containerized or in built enclosures, at the point of use. The user can ramp their own genset up and down as needed. Wastewater treatment plants inject biomethane, and facilities mopping up surplus renewable power might start injecting hydrogen.

Are carbon emissions lower with on-site combustion of natural gas, or by drawing power from the local electrical grid? What is the electrical power grid's projected carbon intensity in 10 years? What if most heat is recovered during generation?

Pipelining natural gas is relied upon by billions of people. It is often preferable to the incumbents (coal or oil) and other viable alternatives (biofuels), so it's widely accepted as a transition technology, provided its methane doesn't leak.

To contain the carbon dioxide from the combustion of gas, post-combustion carbon capture & storage (CCS) is being deployed to intercept carbon dioxide from exhaust rather than release it as emissions. This was touted for *"clean coal"* too but seemingly hasn't met much success there, though pre-combustion carbon capture and storage may fare better. A project underway in the UK intends to permanently bury captured carbon dioxide within rock formations. Another project in Texas uses it, ironically, for *enhanced oil recovery* (OHR). As nobody expects fossil fuel extraction to disappear entirely, it is worthwhile reducing emissions which would otherwise be generated here.

To provide a range of benefits, some tying in with this CCS and proposed at the same sites, that hydrogen economy we discussed is also gradually emerging. This might involve blending hydrogen into the natural gas grid, or converting natural gas pipelines for pure hydrogen service.

## High-pressure gases are dense!

We intuitively think of gas density as being a thousand times lower than liquids – but this is only so at ambient pressure. At a gas pipeline's operating pressure, there are *thousands of tons* of gas in a pipeline. At high pipeline pressures gas densities are of the same order of magnitude as liquid densities. Gas transmission pipelines only became a reality once we found ways to handle these pressures.

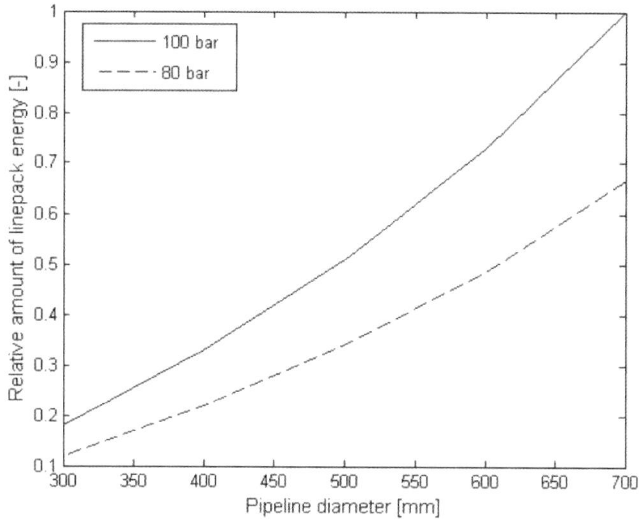


**Figure 95 – Operating pressure can affect a gas pipeline's linepack more than diameter. During operation, as pressure varies depending on flow, so does allowable linepack.** [109]

A gas *transmission* pipeline typically operates above 30 or more atmospheres throughout. It exhibits significant linepack effects, with the total amount of gas contained in the pipeline system varying by a factor of two or more as the system's pressures are raised or lowered in the course of normal operations.

In contrast, an extensive gas *distribution* network might include tens of thousands of short pipes, operating at very low pressures of perhaps a few atmospheres, with negligible linepack. Priorities are different for a transmission pipeline than they are for a distribution network, so a model and a simulator built for one may not suit the other. Atmos SIM handles both at the flick of a switch via a *"distribution or transmission"* toggle, to handle gargantuan networks of 100 000+ pipelines.

# What is it made of? The constituent approach

There are more atoms in a handful of fluid than stars in the observable universe. We need not worry about each and every one, as we are merely concerned with *overall* behavior. The composition of a pipeline fluid for the purposes of pipeline simulation can be considered via one of two approaches.

A fluid mainly consists of pure constituents making up the bulk of it, typically measured in mol%. This constituent approach is the typical way we look at most gas mixtures, whose component breakdowns are well-defined; natural gas pipelines are often equipped with composition meters measuring, in real-time, exactly what the gas is made of at a given instant. It consists primarily of methane ($CH_4$), along with smaller known quantities of other substances.

For a fluid whose component breakdown is known at all times, we can combine the effects of its constituents and calculate their overall properties. Thanks to the meticulous work of physical chemists, how a component interacts with another is well documented. This means that specifying a fluid in this way is generally enough to deduce its behavior: a known composition is enough information for Atmos SIM to calculate a fluid's properties.

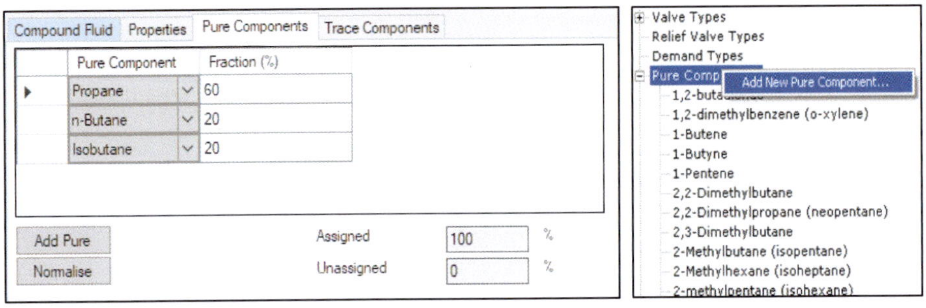


**Figure 96 – A dialogue defining a liquefied petroleum gas (LPG), a pipeline liquid, by its composition breakdown with its constituents listed by fraction (left). Pure components are listed within a default library, and are easily supplemented by adding more as needed (right).**

# Natural gas pipelines

Natural gas is a blend of methane in proportion with other hydrocarbons, and occasionally other gases. Although at atmospheric conditions natural gas is lighter than air, weighing in at less than 1 kg/m$^3$, when natural gas is pipelined at high-pressure it reaches a density of around 275 kg/m$^3$. A batch of, say, liquefied petroleum gas (LPG), despite being a liquid, is only about twice as dense as natural gas at pipeline pressure.

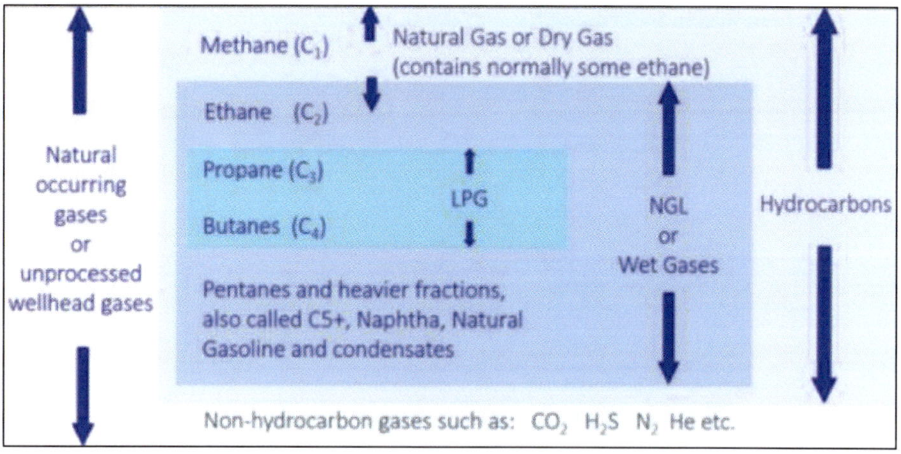


**Figure 97 – Unprocessed wellhead gases is composed of liquefied petroleum gas (LPG), natural gas liquids (NGL) and other constituents (Credit: NorwegianPetroleum.no)**

---

How much denser is natural gas in a high-pressure pipeline (190 barg, 20°C) than at atmospheric pressure?

Methane generates more useful energy per unit of carbon dioxide emitted than any other carbon-based source. At the same time, methane also happens to be the only fossil fuel which is *itself* a greenhouse gas: methane's greenhouse effect is far more potent than carbon dioxide, and it is responsible for a quarter of man-made global warming. [40] Methane emitters include landfills, whose gas can be captured for combustion in special engines on site. Cattle manure is another emitter. The natural gas industry itself accounts for around 10% of methane emissions. [39] Detecting methane leaks from equipment throughout our natural gas infrastructure is therefore a major objective for society at large.

Disappointingly, in 2019, we flared and even vented more natural gas than ever before. The US wasted over 1% of its gross gas withdrawals. Where *venting* is still allowed, we are failing to avert the serious harm of methane's hideous greenhouse effect. Wasting gas for no tangible use is the worst outcome for all. In 2019, the world wasted an incredible 150 billion cubic meters of gas. That could easily have heated 80 million homes! Left unchecked, our trajectory is getting worse. Policymakers are obliged to step in. The World Bank aims to end both of these wasteful practices through its *"Zero Routine Flaring by 2030"* initiative. One attributed cause was poor pipeline and processing infrastructure around oilfields, such as in Texas and North Dakota. [41] Gas that is *"stranded"* can only be *"rescued"* if pipelines can transport it to processing, and on to viable markets.

Upstream, natural gas might be wet, containing significant humidity: water content. It must be kept warm enough to stay in its vapor state, to avoid forming dangerous fast-moving liquid droplets. A natural gas pipeline is routinely equipped with dehydration stations and measures in place to remove this humidity. If incoming gas has been pipelined a long way, it is often quite dry. Natural gas can also be classified as sour or sweet based on its hydrogen sulfide and carbon dioxide content, and rich or lean based on condensates. We take another closer look at natural gas composition in the next section of this chapter.

The demand for natural gas has driven innovations that transport it across the world in ships, giant tankers at refrigerated temperatures, in the form of Liquefied Natural Gas (LNG), to receiving terminals linked to local pipelines. This has created a market where it is traded as a commodity, bought and sold based on its combustion value. It's replaced liquid fuels in many applications.

The proportion of natural gas users varies country-to country. Numerous industries rely on methane as a feedstock, including for production of ammonia, hydrogen and methanol. A major demand is for domestic boilers and cookers, replacing unwieldy propane canisters for example, and relied upon for heating in cold climates, such as northern Europe. Power generation is also a major user, with combined cycle gas turbine (CCGT) power plants proving favorable to fund and responsive enough to stabilize the electric grid.

This is a far cry from its humble origins. Natural gas was originally flared off: burned in a tall chimney stack, intentionally wasted as a byproduct. As the 1900's progressed, advanced processing and pipelining supplied users with plentiful natural gas. It displaced its predecessor: dirtier-burning towns gas, which was distributed in a dedicated low-pressure grid, one for every town. It is also called coal gas, as it was manufactured from coal, at great expense. This gas had poor energy content which was a hindrance. Interestingly, it was partly made of hydrogen. By this quirk of history, old gas grids such as that in the UK, and all appliances connected to it, are to some extent already capable of handling a certain proportion of hydrogen. [42] Trials to blend hydrogen into the grid were a success. Gas transmission pipelines are working with suppliers and customers to accommodate more hydrogen within quality specifications.

Special gases is a blanket category referring to many kinds of combustible gases other than natural gas. They are classified according to their heating value. A rule of thumb is that gases whose heating value exceeds 33.5 $MJ/m^3$ are amenable to pipelining. These consist mainly of methane, and are referred to as substitute natural gas (SNG). Blending biogas from wastewater treatment plants, as well as hydrogen and other special gas streams (e.g. landfill gas, after cleaning it) is case-by-case. It depends on local regulations as well as pipeliners' specification of what they can tolerate.

Natural gas is a mixture of pure components. Calculations determine whether a natural gas is acceptable for delivery based on its composition at any given place and time. Parameters which determine this include the *incomplete combustion factor* (ICF), *soot index* (SI) and *Wobbe index* (WI). A carbon number is used to translate an alkane, alkene or aromatic component present in natural gas to an equivalent combination of methane and propane, in terms of number of carbon and hydrogen atoms.

# Constituents of natural gas

Gas processing facilities make a natural gas *pipeline-ready*. Part of this aims to avoid hydrate formation. Removal of water vapor (*humidity*) is achieved by a process dehydrating *wet gas* to make *dry gas*. This is typically reinforced at some stations along a gas pipeline, which might also be equipped with drums to *knock-out* any condensate.

Condensables are heavy hydrocarbons, undesirable in a gas pipeline because they might *condense* and drop-out along its length, as *condensate*. A gas containing a lot of these is a *rich gas* (over 668 $cm^3/m^3$). If it doesn't it's a *lean gas* (under 334 $cm^3/m^3$). Once they have been recovered, these are *natural gas liquids* (NGL's), refined to *gasoline-range* (C6+) liquid products. A pipeline carrying rich gas may see *liquid slugs*. As discussed in the main book, this requires simulators to go *"Beyond single phase"*.

A third task for gas processing facilities is to remove acid gases. A chief concern is *hydrogen sulfide*, corrosive and toxic. $H_2S$ content controlled within 5.7 $mg/m^3$. Another is carbon dioxide. $CO_2$ content may be tolerated within 3 mol%. Refinery processes scrub and absorb to perform the sweetening of *sour gas* into *sweet gas* use hydrogen. This application is a prime contender for an emerging low-carbon hydrogen economy.

And finally, a natural gas might also be *diluted* by the presence of inert gases (*inerts*). Significant proportions of nitrogen, for example, are detrimental to the calorific value of a natural gas. At best, energy expended transmitting these inerts is essentially wasted.

These terminologies all take some getting used to. Unprocessed gas from some production fields might already be sweet. Some processed gases are still wet. To further complicate matters, there seems to be some inconsistency in the terminology chosen by the various gas pipeliners as well as wider teams within the gas industry.

There is no single universal *"pipeline-quality gas"* to speak of. Because the processing plant removed ethane and higher alkanes, heating value might need boosting before a gas is accepted by customers. The European Gas Research Group (GERG) classify pipeline gases by composition into six groups, commonly referred to after gas fields.

| Component \ Group | I | II | III | IV | V | VI |
|---|---|---|---|---|---|---|
| Methane $CH_4$ | 81.21 | 89.45 | 84.47 | 57.69 | 85.17 | 98.27 |
| Ethane $C_2H_6$ | 2.81 | 0.24 | 8.86 | 1.78 | 5.40 | 0.52 |
| Propane $C_3H_8$ | 0.38 | 0.02 | 3.18 | 3.2 | 1.59 | 0.16 |
| Butanes $C_4H_{10}$ | 0.13 | 0.01 | 0.97 | 0.15 | 0.51 | 0.06 |
| Pentanes $C_5H_{12}$ | 0.05 | 0 | 0.20 | 0.04 | 0.15 | 0.02 |
| Higher hydrocarbons $C_6$+ | 0 | 0 | 0 | 0 | 0 | 0 |
| Hydrogen $H_2$ | 0 | 0 | 0 | 35.63 | 0 | 0 |
| Carbon monoxide CO | 0 | 0 | 0 | 0 | 0 | 0 |
| Carbon dioxide $CO_2$ | 0.99 | 4.86 | 1.85 | 0.03 | 1.46 | 0.07 |
| Nitrogen $N_2$ | 14.38 | 5.37 | 0.43 | 1.42 | 5.68 | 0.89 |
| Helium | 0 | 0.05 | 0 | 0 | 0 | 0.02 |

**Table 6 – Composition in mol% of GERG's representative natural gases [43]**

- Group I; nitrogen >0.095 mol% and no hydrogen, such as Groningen gas
- Group II: carbon dioxide >0.04 mol%, such as a Drohne gas
- Group III: ethane >0.08 mol%, such as an Ekofisk gas
- Group IV: hydrogen >0.02 mol%, such as a Bacton gas
- Group V: nitrogen and ethane each >0.04% and no hydrogen: TENP gas
- Group VI: methane >0.94 mol%, such as a Russian pipeline gas

American organizations such as NIST make their own classifications, named after gas fields. Amarillo represents moderate gas, and other fields represent various richer and leaner gases much in the same way as shown for GERG. [44]

# Hydrogen in pipelines

Blending hydrogen into the natural gas grid and the pipelining of hydrogen is a fascinating technical topic. We shall take the time to discuss it in its own appendix.

# Nitrogen in pipelines

When commissioning a gas pipeline, purging with an inert is done at around 2 barg, flowing at a high enough velocity to clear accumulated dust (typically 15 m/s). Each segment is then typically pressurized to pack around 10 barg. [75]

Whenever a pipeline needs to be worked on by welders or others, such a purge needs to happen again. Any flammable fluid within it must be swept away. In the 1950's, people used compressed air to do this. While air was sweeping out crude oil, disaster struck. Detonations occurred along a whopping 30 mile stretch of the crude oil pipeline. Kletz was not allowed to disclose where!

Today, the industry knows better, and uses inert nitrogen. Even following a nitrogen purging procedure, liquid might potentially remain at the low-points, so drain points are fitted strategically along a pipeline to cope at such locations.

Nitrogen is also used in some gas pipelines to dilute natural gas in order to meet standards of quality and heating value. This poses a particular challenge for some pipeline simulators, as if they neglect to recognize that *"flavors of flow rate"* are non-interchangeable, they might end up describing flows only in terms of energy content. This makes flowing nitrogen an *invisible stream*. In fact, even non-primitive pipeline simulators might overlook a pitfall of dividing by zero somewhere if they attempt a calculation involving zero-energy-content.

# D7 – ENERGY VALUE AND ENERGY DENSITY

Some of the heat released by combustion of a hydrocarbon ($C_mH_n$) is used to immediately vaporize the water ($H_2O$) formed during the combustion:

$$C_mH_n + O_2 \longrightarrow \frac{n}{2} \cdot H_2O + x \cdot CO_2 + y \cdot CO$$

What the user pays for is called by many terms, referring to the same quantity:

- Gross calorific value (GCV)
- Higher calorific value (HCV)
- Higher heating value (HHV)
- Higher calorific energy (HCE)

The user only makes use of *available* energy, excluding a portion of the total energy immediately consumed to vaporize the water formed in the reaction. This available energy is much lower, and goes by any of the following names:

- Net calorific value (NCV)
- Lower calorific value (LCV)
- Lower heating value (LHV)
- Lower calorific energy (LCE)

If you use another name for these, just write in, and we can add it to the list.

Can a high-pressure pipeline transport more combustion value in the form of a liquid fuel, or as natural gas?

Fossil fuels are spoken about in terms of their combustion value as this is their primary use, but one mustn't forget that these same substances also serve as essential feedstocks for all kinds of petrochemical and manufacturing industries. A common grounds for comparing different fossil fuels is by expressing their quantities in terms of *million tons of oil equivalent* (MTOE) – a frighteningly large amount, reminding us of the sheer scale of the industries.

Liquids offer an energy density whose order of magnitude is unmatched by any compressed gas or battery technology. But the energy density of a fuel is only one factor amongst many to bear in mind. It doesn't always matter. To weigh up what options work best for a given application, an engineer also considers:

- Safety – battery chemistries vary drastically in this regard
- Cost and availability of materials – platinum, needless to say, is rare
- Efficiency – usually converting some kind of potential to useful motion

This means that what works best for vehicles in urban Brazil isn't necessarily what works best for off-grid stationary generation in an isolated part of Russia. In the most remote locations, shipping a fuel in accounts for most of its price. Where a major industry like mining dominates a local economy, bulk-buying fuel saves all residents money. Electrifying those industries can potentially help or hinder surrounding communities depending on precisely how it's done!

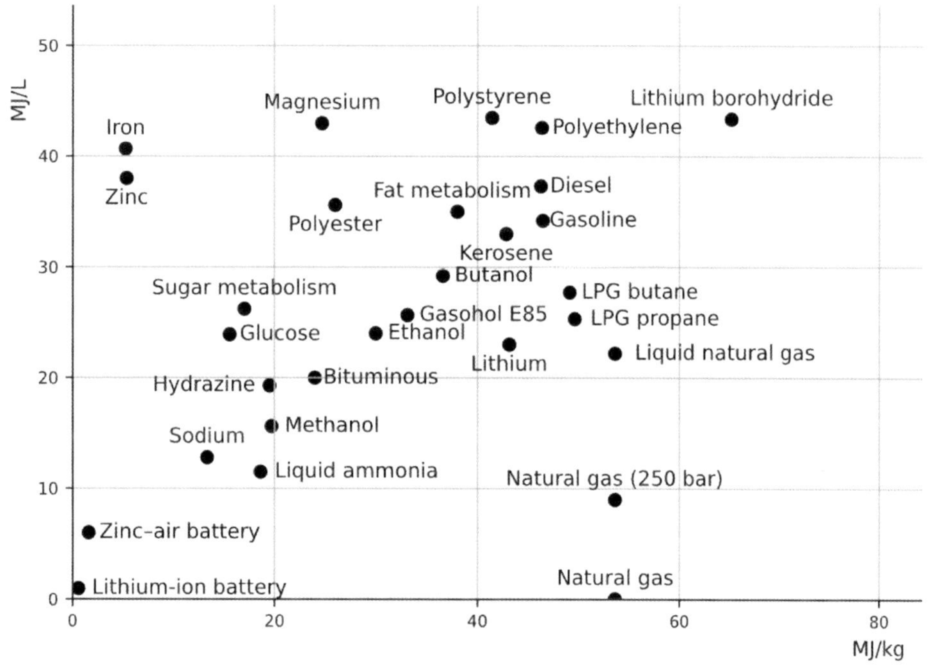


Figure 98 – Energy density by volume and by mass of various batteries and fluids

| Storage technology | Principle | Energy density | |
|---|---|---|---|
| | | kJ/kg | kJ/L |
| **Pumped hydro (200 meters)** | Mechanical | 1 | 1 |
| **Supercapacitor** | Electrical | 50 | 30 |
| **Solar salt (275 to 550°C)** [45] | Thermal | 1 200 | 675 |
| **Lithium-ion battery** | Chemical | 950 | 2 400 |
| **Fuel** | **State** | **Combustion Value** | |
| | | kJ/kg | kJ/L |
| **Ethanol or bioethanol** | Liquid | 26 000 | 21 000 |
| **Gasoline / petrol** | Liquid | 43 000 | 32 000 |
| **Biodiesel** | Liquid | 37 000 | 33 000 |
| **Diesel** | Liquid | 42 000 | 36 000 |
| **Liquified petroleum gas (LPG)** | Liquefied | 45 000 | 24 000 |
| **Liquefied natural gas (LNG) at −160 °C** | Liquefied | 53 000 | 22 000 |
| **Compressed natural gas (CNG) at 250 barg** | Supercritical | 53 000 | 9 000 |
| **Liquid hydrogen at −252 °C** | Liquefied | 120 000 | 8 500 |
| **Hydrogen at 300 barg** | Supercritical | 120 000 | 2 500 |
| **Hydrogen at 800 barg** | Supercritical | 120 000 | 5 200 |

**Table 7 – Energy densities of technologies and combustion values of substances expressed by mass (kJ/kg) and by volume (kJ/L)**

This basic comparison isn't necessarily like-for-like. The energy industry has defined a *levelized cost of storage* (LCOS) and a *levelized cost of energy* (LCOE) indicating the equivalent prices. Storage technologies must consider *power* density along with energy density. And we must ask what the full-cycle efficiency is for *long-term* storage. Today, it takes too much energy to produce hydrogen, compress it for storage, and convert it into electricity in a fuel cell. These factors conspire to make today's hydrogen too expensive to be a versatile energy carrier.

# APPENDIX E – HYDROGEN ALTERCATIONS

*"Water will be the coal of the future"*

– Jules Verne, 1874

## E1 – ENERGY STORAGE AND HYDROGEN

Hydrogen is in certain places no longer merely a specialty chemical supplying our industries, but being championed as an emerging energy carrier to plug a technology gap in grid-scale inter-seasonal energy storage. Methanol and other fluids are also being touted for the same purpose, with sceptics objecting to it as hype. To be deployed at scale, technical and commercial challenges must be overcome.

There is heated and very public opposition to the hydrogen economy, the role it is already beginning to play in several countries, and its expected impact. Famously, Elon Musk called using *"fool cells"* for vehicles *"mind-bogglingly stupid"*. The biggest stone hurled at hydrogen's glasshouse is that it's always inevitably going to be expensive, as energy is inevitably wasted to generate it. Mention hydrogen, and we'll be shot down, reminded of the merits of electricity. A palpable look of fear may appear in the eyes of the fossil fuel worker. We're led into a lengthy separate discussion on energy storage.

One argument *for* hydrogen goes something like this. Electrical grids in 2020 are having to resort to curtailment, throwing power away. People seem to dispute this, but the UK paid one of its few nuclear reactors to switch itself off recently, and if that isn't curtailment I don't know what is. To draw this power on demand, mopping up surplus spikes from renewables, desalination of seawater is one option. But that comes with its own environmental and economic downsides. Besides, not everywhere needs more water – we have quite enough in the rainy north of England.

Lithium-ion batteries are only capable of effectively storing power at scale for a few hours at a time. I am no expert on their materials, but a lot of people would make a lot of money notching them up by orders of magnitude, in terms of both energy stored and duration. Yet nobody is optimistic that it's possible. Hiccups aside, they were never designed or intended for stationary application. There are high-profile news stories about grid-scale battery storage in places such as Australia and California, and indeed it is a useful technology, but the capacity being installed on a scale of gigawatts, and the world needs terawatts.

Alternative chemistries do exist. A promising one is liquid-metal, hoping to be cheap and scalable by avoiding rare metals. It, and others, are enthusiastically pursued by start-ups, but all remain quite some way off mass manufacturing. This is particularly worrisome as they must operate at incredible temperatures, and it takes a while to scale-up any technology, proving it's affordable and safe.

A rare, obvious way of using surplus power at our disposal is to make green electrolytic hydrogen. Hydrogen is amenable to long-term storage in large quantities, pressurized or liquefied in vessels, or in underground salt caverns. However, this rosy picture does not yet reflect reality. Only a minuscule proportion of today's hydrogen is *"green hydrogen"* produced via the expensive electrolysis of water using electricity. And what is the point of using hydrogen if it was generated from fossil fuels in the first place?

We have hydrogen today, but not cheap electrolytic hydrogen. This is the missing piece of the puzzle – well it's more like a *big gaping hole* in the puzzle. Electrolysis is still unaffordable at time of writing. Many governments and companies are investing substantially to make it cheaper, but we're not there. The price that it needs to reach to become competitive is reportedly something like a couple of USD per kilogram, with maybe fifty cents tacked on to ship it. When Brits say something sounds *optimistic*, they do so in a sarcastic sense. Yet it does genuinely seem people are optimistic that economics of scale are about to make hydrogen affordable, particularly if carbon taxes unlock *"blue hydrogen"* – which is generated via conventional processes, steam reformation of natural gas or coal gasification, but coupled with carbon capture so as have much lower greenhouse gas emissions. Could this help build up infrastructure while cheap new electrolyzers are developed to mop up offshore-wind power, or supply chains connect the farthest reaches of sunny deserts to industries?

**Figure 99 – Offshore wind turbines are generating renewable electricity at an immense scale**

# E2 – WHO MAKES HYDROGEN?

## Hydrogen's true colors

Although hydrogen is the most abundant element in the universe, it doesn't naturally occur in its pure form. Making clean hydrogen economically is hard. Ideas abound, ranging from likely to tentative to outlandish. *"Yellow hydrogen"* would be made using heat raised by nuclear reactors. *"Turquoise hydrogen"* would be made via pyrolysis of methane, yielding solid carbon. And so forth.

Some are working on *biomass* to hydrogen, not to be conflated with anaerobic digesters. Rather, biohydrogen would be produced by *gasification* of biomass, followed by water-shift reactions to hydrogen. It comprises the first stages of a longer gasification process already in use to produce synthetic natural gas (bio-SNG). I am not quite sure what color that would be, if it works out.

In 2019, an IEA report (*The Future of Hydrogen*) concluded that despite false starts in the past, this time the momentum could be different. Just as low-carbon technologies saw unprecedented success scaling up solar panels, wind turbines and lithium-ion batteries, electrolyzers may soon become affordable. That electrolytic process is called *"green hydrogen"*. It is hoped that projects being deployed to make low-carbon *"blue hydrogen"* by carbon capture will help by satisfying existing users' demands, as well as building up necessary system infrastructure and driving market scale. [101]

| | |
|---|---|
| Grey hydrogen | Manufactured from natural gas, incurring more emissions if used for energy than natural gas itself. |
| Blue hydrogen | Manufactured from natural gas, with carbon capture and storage during the manufacturing process. |
| Green hydrogen | Manufactured by electrolysis powered by renewable electricity, generated by wind power or solar power. |

**Table 8 – Working definitions of grey hydrogen, blue hydrogen and green hydrogen**

# Grey hydrogen: methane reformers or coal gasifiers

As it stands, 95% of hydrogen is made from fossil fuels, with the other 5% generated as an industrial by-product in the chlor-alkali process. * The *"grey hydrogen"* produced from methane is generated by a process known as *steam methane reforming* (SMR). For every unit of methane it consumes, it produces a quantity of hydrogen containing less energy than went into it; this is a feature of any industrial process and is a limitation imposed by the fundamental physics of thermodynamics. It supplies the major existing users of hydrogen, mainly refineries and other industries. [102] Steam-methane reforming (SMR) breaks methane in natural gas with steam and *"re-forms"* it into hydrogen.

Steam *"re-forms"* methane into syngas $CH_4 + H_2O \rightarrow 3\,H_2 + CO$

More steam *"shifts"* syngas into hydrogen $CO + H_2O \rightarrow H_2 + CO_2$

The first step is the reforming, and the second is the water gas shift reaction.

Overall, both steps result in this balance $CH_4 + 2\,H_2O \rightarrow 4\,H_2 + CO_2$

Coal is converted to hydrogen by a more involved set of reactions within a *gasification* process, one that also releases carbon dioxide emissions. For this reason, the world consumes 6% of its natural gas and 2% of its coal making this grey hydrogen, emitting more $CO_2$ than the UK + Indonesia combined.

**Figure 100 – Coal (foreground, unloaded at a harbour) and natural gas are already being used to manufacture hydrogen which is in demand for various chemical industries (background)**

---

* From the electrolysis of brine, related but different to dedicated electrolysers.

# The trouble with post-combustion carbon capture

Methane combustion at point-of-use leaves us with one option that has been around for years. This is *post*-combustion carbon capture, and it is expensive:

- Carbon dioxide is dilute in exhaust: $CH_4 + O_2 + N_2 \rightarrow H_2O + CO_2 + O_2 + N_2$
- A chilled solvent absorbs that carbon dioxide: $NH_3 + H_2O + CO_2 \rightarrow NH_4HCO_3$
- Steam regenerates loaded solvent: $NH_4HCO_3 + H_2O \rightarrow (NH_3 + H_2O) + (CO_2)$

The example given here uses ammonia, similar to the amine used in practice. Many solvents work similarly. The aim is to purify $CO_2$ so it can be pipelined.

Although we used single-headed arrows, in reality all of these equilibrium reactions are happening in both directions, so it is particularly hard to purify carbon dioxide from an incoming stream in which it is very dilute (a few percent) and at low pressure to boot.

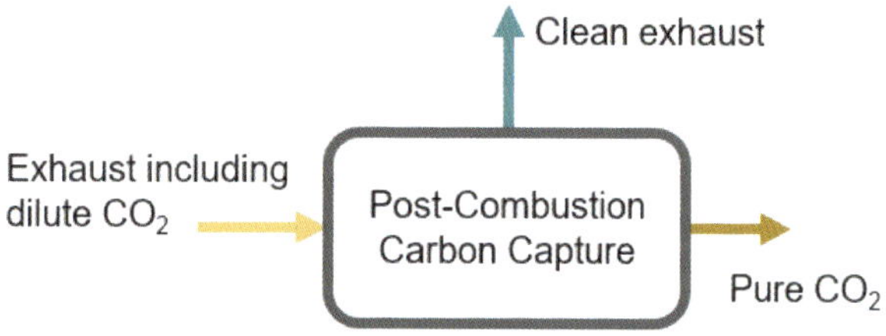


Figure 101 – Schematic of inflow and outflows for *post*-combustion carbon capture

This post-combustion carbon capture process is expensive for several reasons. Energy is consumed to attain conditions that keep these reactions happening effectively – refrigerating the absorber, heating and pressurizing a regenerator, circulating the solvent, etc. There are also chemicals consumed in the process; an acid wash column prevents solvent slipping into the exhaust.

Moreover, every exhaust needs its own new $CO_2$ pipeline to long-term storage, so although this approach has been deployed successfully at some major emitters such as refineries and gas-fired power plants, all those limiting factors have conspired against post-combustion carbon capture. It only gets us so far.

# Blue hydrogen: auto-thermal reformers

Low-carbon hydrogen is already within reach using *carbon capture*. We can't just retrofit an existing grey hydrogen plant by placing a clean-up unit onto the end of its exhaust and then expect to curb its emissions. Even if it is located near a suitable geological site for storage, existing steam-methane reforming plants are poorly suited to carbon capture technology. We can capture no more than 80% of $CO_2$ by retrofitting, due to the relatively low pressure of the $CO_2$.

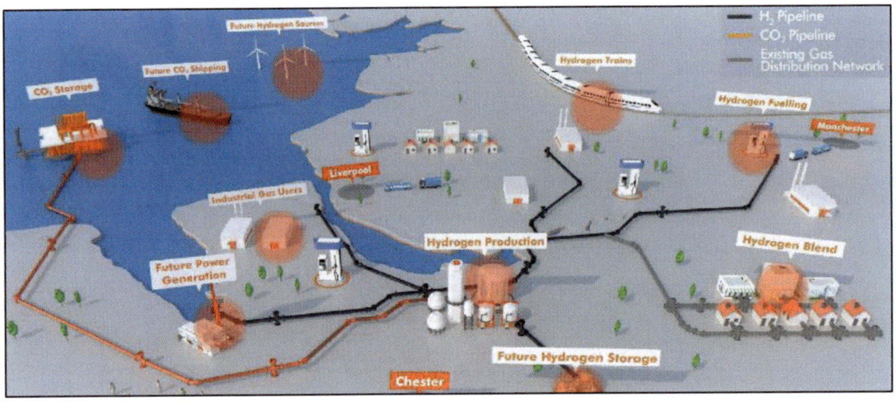


**Figure 102 – A project to produce blue hydrogen, capturing carbon and storing it under Liverpool bay, supplying low-carbon hydrogen to a diverse user base (Credit: the HYNET carbon capture and storage project in North West England)**

The lowest hanging fruit for cleaner *"blue hydrogen"* capturing 97% of $CO_2$ is the development of dedicated *auto-thermal reformation* (ATR). That process is what is amenable to *pre*-combustion carbon capture and storage (CCS). The processes are integrated from the outset during the design phase. It produces carbon dioxide in a form concentrated to 50%, and at substantial pressure – conditions favorable for the purification processes *"capturing"* it downstream. Designing our process so the carbon dioxide is easy to capture from hydrogen is what allows for that higher rate of capture. Even better, the dedicated design improves the conversion of methane to hydrogen. Producing 320 MW worth of hydrogen in this way takes just 400 MW of natural gas, instead of 440 MW. [103]

Siting the facility appropriately lets us pipeline the $CO_2$ to nearby geological stores. The 60 km carbon dioxide pipeline of the HYNET project plans to initially transport $CO_2$ in the gas-phase on land. In the initial stages of the project, it will also be flowing in the gas-phase offshore. As additional $CO_2$ capture projects are brought online, it can transition to flow in the dense phase. Both on-shore and off-shore portions consist of pre-existing pipelines that are to be repurposed for this new service. [104]

This is the basic structure of all the hydrogen deployment projects across the UK. But what about elsewhere? In 2018, the IEA estimated blue hydrogen made via CCS in the US, Middle East or Russia to be cheaper than even grey hydrogen made in Europe or China. However, one should be careful in defining how clean the process used will be.

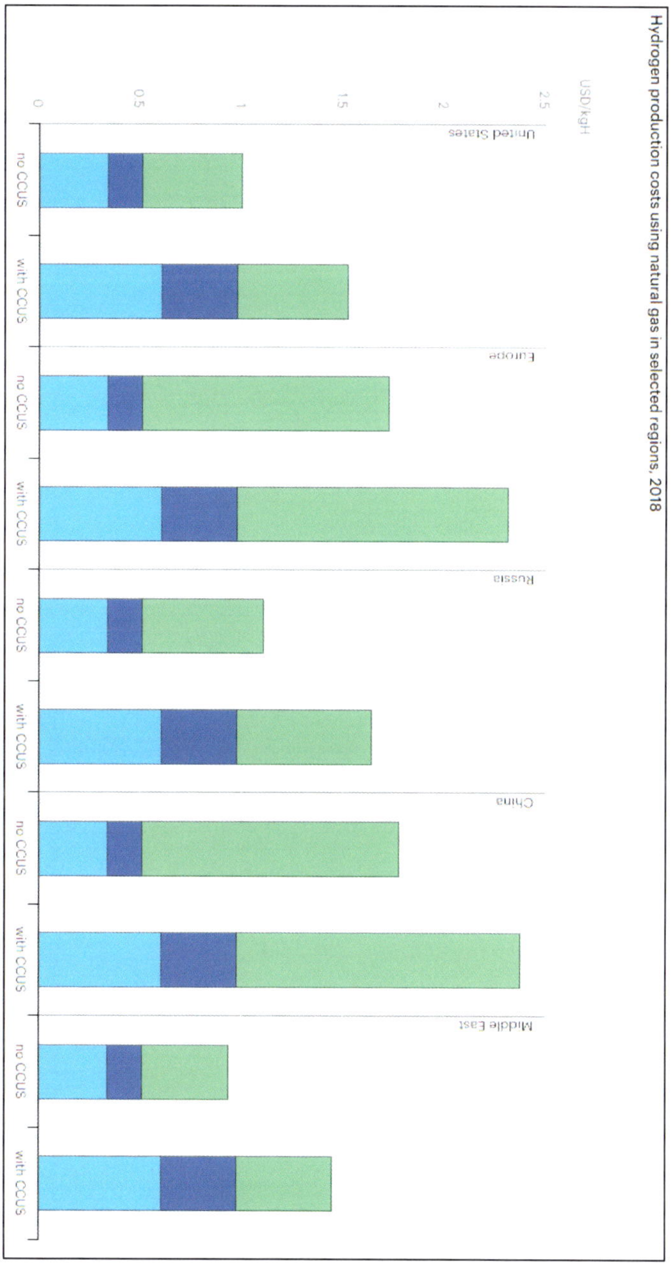


**Figure 103 – Low-carbon hydrogen made in the US, Russia or the Middle East is thought to be cheaper than producing hydrogen even by polluting methods in Europe and China (Credit: IEA)**

# Green hydrogen from water using electrolyzers

The barrier for electrolyzers to make power-to-gas facilities a reality on an industrial scale is not technical; it is the cost of the hydrogen they produce. That is addressed by scaling up its production and choosing the best locations. The world energy council predicts a possible hydrogen market of 270 million tons by 2050. If this were made using green hydrogen electrolyzers alone, it would consume all of today's globally installed capacity for renewable energy. That is not an insurmountable obstacle; the world has already started ramping up its deployment of renewable power generation. The big question is *where*.

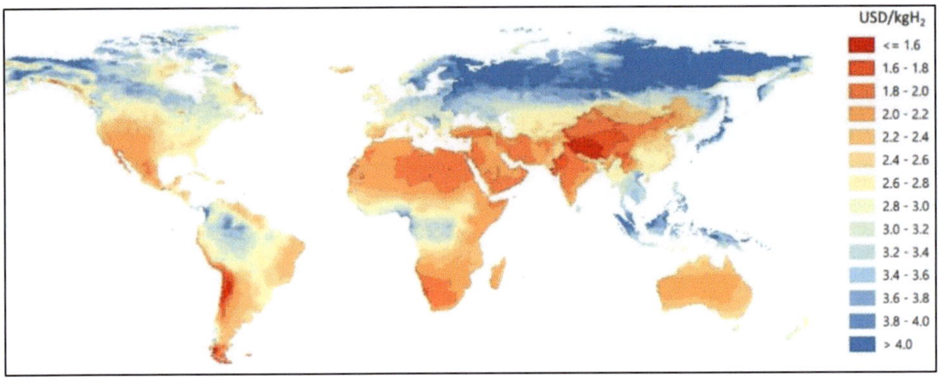


**Figure 104 – Electrolytic hydrogen is cheaper to manufacture in remote places (Credit: IEA)**

The answer to that question is often remote locations. So how do we transport cheaply produced electrolytic hydrogen from the sunniest, windiest parts of the world to major centers of population and industry? It may sound outlandish to suggest liquefying or compressing it and shipping it across sea and ocean routes, but that is exactly what the energy market achieved already with liquefied natural gas (LNG). Transportation becomes a critical part of the hydrogen value chain, and pipeliners should realize that our pipeline infrastructure in this context could become especially significant.

In Europe, the push is to make hydrogen using the suddenly ample power generated by offshore wind. This would involve transmitting the power to onshore electrolyzers via high-voltage cables, or locating the electrolyzers offshore. An offshore electrolyzer would preferably process seawater into hydrogen, rather than having to pipeline water all the way out to the middle of the North Sea!

# E3 – WHO USES HYDROGEN?

## Hydrogen for transport

Figure 105 – A hydrogen-powered train (left) and the Mirai, a hydrogen fuel cell vehicle (right)
Michael Kramer / Wikimedia Commons / CC BY-SA 3.0

What we use hydrogen for is its own arena for fans of loud, raging debate. The only talking point one seems to hear with regards to hydrogen is the day's fashionable topic: decarbonizing transport. People hear that and think *electric cars* rather than trains or trucks, and so unfairly dismiss hydrogen out of hand. Hydrogen for transport is not straightforward. Distance, torque or uptime may favor fuel cell electric vehicles (FCEV), so serious enterprises are looking into it. But as of the time of writing, Scania – a global major manufacturer – cut short its trials, citing the prohibitive costs of inefficiency, repair and maintenance compared to batteries. So is the same conclusion valid everywhere? Can hydrogen-powered trains carry enough fuel and generate enough power to wean us away from diesel locomotives on hard-to-electrify rail routes? Many open questions remain unanswered – this is a space to watch as trials and research are still ongoing.

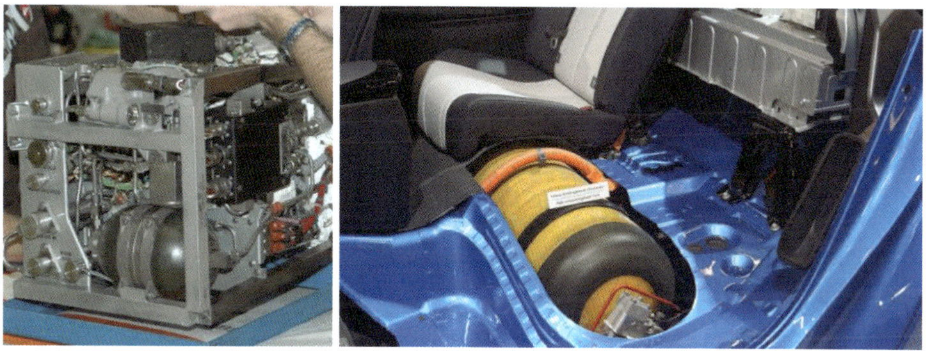

Figure 106 – A hydrogen fuel cell (left), and a hydrogen tank and electric battery in a car (right)
Mario Roberto Durán Ortiz / Wikimedia Commons / CC BY-SA 4.0

# Hydrogen for heating

If too many homes in cold climates are converted to be heated by electricity rather than natural gas, an electrical grid might become unable to cope with the magnitude of changes in supply and demand. Could hydrogen supplement the natural gas grid, so as to partially decarbonize this existing infrastructure? Germany discounted hydrogen for domestic heating entirely. But the UK plans to complement the electrification of heating with hydrogen-for-heating projects, improving insulation (at time of writing, still taxed higher than fossil fuels!) and ground-source heat pumps. The UK's stated goal is to heat 15% of homes with low-carbon hydrogen, blended or pure. Existing natural gas grids are being upgraded, cast iron pipes replaced with hydrogen-ready plastic. The reason for this difference in approach might relate to the nature of homes in each country, or the cost of heat pumps at a location compared to the cost of making greener gas. It may also reflect the difference in the UK's industrial base compared to Germany. Both country's policymakers appear committed to their own stance.

It is easy to blend hydrogen into natural gas. Though it has poor combustion properties, a hydrogen + natural gas blend can handled fine by ordinary domestic boilers and cookers. The pipelines are there to transmit and distribute it, and can be adapted to be suitably specified without disruption. Combusting a blend in *"normal"* equipment avoids convincing or coercing consumers to pay for significant change. People don't like disruption and don't want to incur extra costs – though there is a premium most are willing to accept, to play their part in decarbonizing. Their acceptance and trust is instrumental when someone proposes pipelining a new gas into their home!

Small scale trials blending hydrogen to some UK homes have succeeded, with customers experiencing no noticeable difference with existing equipment. As hydrogen for heating homes is rolled out to enable blending up to 20% into the gas grid in Leeds, detailed studies are underway. Would leakage be any different to natural gas? Would the consequence of a release in buildings, or how we do operations, maintenance and repairs, be any different with hydrogen? Can gas suppliers continue to be 99.99% resilient? [105]

Delivering a blend ever richer in hydrogen hits a limit, beyond which it makes more sense to carry pure hydrogen. This is because hydrogen burns in what is akin to an explosion, so although it is tolerable up to a point in a natural gas blend, there is a limit. When combustion stops making sense, so do boilers.

# Hydrogen for electricity

Although some refineries burn syngas in turbines for electricity, converting *pure* hydrogen to generate electricity and heat makes sense using a fuel cell. Japan already has hundreds of thousands of natural gas fuel cells, as micro combined heat and power systems (micro-CHP). In the aftermath of the Fukushima disaster, Japan became the world's most serious pursuer of hydrogen in all its uses; stationary CHP, industries, and yes even hydrogen fuel-cell electric vehicles (HFCEV). Their peculiar situation, shipping in much of their energy as liquefied natural gas rather than pipeline, has meant they perhaps have the world's only *negative spark gap*. Incentives were applied such that it *never pays* to burn natural gas for electricity in Japan. Everywhere else in the world, we pay people to sit ready to burn natural gas! At some peak times, renewables fail us. When they aren't generating enough electricity, so we resort to paying peaker plants to burn fossil fuels, bailing us out with electricity generated from gas. Storing electricity is currently impossible at a large enough scale, for long enough periods.

Generating so much clean hydrogen that it goes from being a feedstock to becoming an energy carrier may allow us to use it as a form of energy storage. This use case – hydrogen for power – is championed by some as a way to help relieve the electrical grid. It is criticized for being inefficient, yet it can't be any worse than our present approach, subsidizing carbon-emitting plants to bail us out in our times of need, while throwing renewables away at times of plenty. It appears government strategists have decided this is a more likely hurdle to overcome than waiting for new batteries capable of grid-scale inter-seasonal storage to go into cheap mass-production, though the mechanisms are in place to accommodate them if they do somehow materialize. The official position in many countries envisages hydrogen will be a new tool in our arsenal, for electrical grid operators to perform seasonal balancing. Strategies diverge on using hydrogen to heat homes or generate electricity, but it is already being used in industry.

There is no shortage of detractors furiously opposing hydrogen. Anyone involved with it should be prepared for the storm. Detractors condemn it as a mere ploy by the usual suspects: oil and gas majors, plotting and scheming to get their hands on a new wave of lucrative government subsidies. What about the limit we're hitting with renewable generation? New ways of storing enough electricity long-term will appear, they insist.

It's inefficient to generate green electrolytic hydrogen, but getting *something* from power that is otherwise wasted – even if the reader is ambivalent towards natural gas and opposed to any-color hydrogen that isn't essential – is surely worthwhile. And it's massively expensive to build carbon capture and storage, but cleaning up our act where we already generate grey hydrogen from fossil fuels is a practical responsibility.

# Hydrogen for industry

Industries and refineries already manufacture dirty *"grey hydrogen"*, emitting carbon in their existing processes. Renewables are oftentimes generating surplus electricity, forcing us to throw that power away by curtailment. Those two problems might support one another if electrolyzers could be scaled up to turn electricity into hydrogen.

Acknowledging that some applications *can't be electrified* is a fact that's worryingly absent from the public psyche. This is a shame: in many applications, vital industries are stuck with feedstocks or energy dense fuels for the foreseeable future. Where electrification is impossible, hydrogen is often front-and-center. There are glassmaking furnaces, brick kilns, and large industrial steam and hot water boilers. Some already use hydrogen directly, while others can switch to using it instead of methane. Some rely on furnaces, which can be adapted to use hydrogen either in its pure form or as *"syngas"*, but otherwise fail to reach sufficient temperatures if *"electrified"*. Industrial clusters are the top priority for many policy-makers: industry needs hydrogen.

In light of that fact, hydrogen for electricity seems like a pipe dream. Can this feedstock, today merely an industrial gas, not only generate heat but power?!

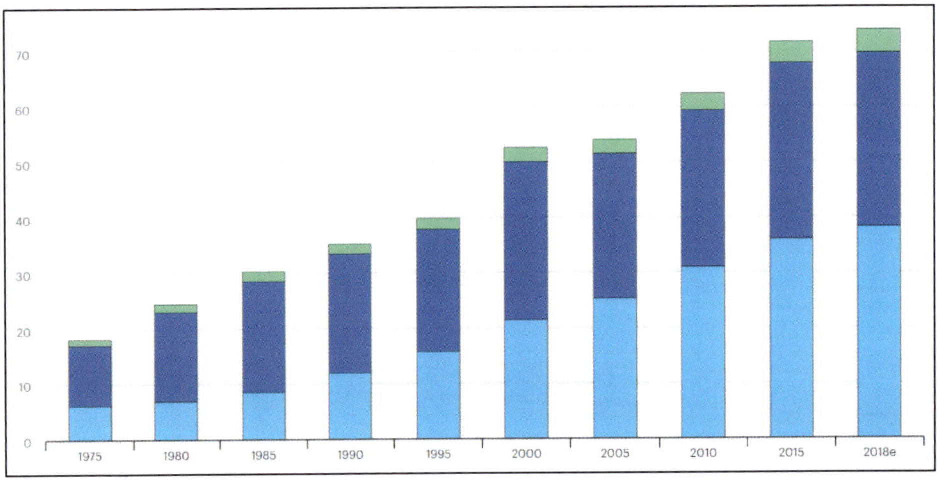


**Figure 107 – Worldwide hydrogen demand for refining (cyan) and chemicals (blue) is an order of magnitude greater than that used for everything else (green) (Source: IEA)**

Production of electrolytic hydrogen stands at a measly rounding error. Before low-carbon or zero-carbon hydrogen can become the versatile energy carrier advocates envisage, it must be produced in sufficient quantities to satisfy that existing demand. Most of the world already manufactures hydrogen for its industries. In 2018 those existing users constituted around 70 megatons of hydrogen consumption as an industrial gas, mostly for refineries and factories producing ammonia, methanol or steel. Producing all of today's dedicated hydrogen output from electricity would represent a demand of 3 600 TWh. This exceeds the total electricity generated annually in the European Union. [101]

# What questions should we be asking?

In brief, the world is facing several problems whose solutions might be linked. Cleaning up today's hydrogen using carbon capture and storage (CCS), and anticipating the emergence of cheap plentiful electrolyzers, inevitably raises questions about transporting hydrogen in new ways. Those questions stand regardless of whether we continue using that hydrogen only as an industrial gas, or start using it as an energy carrier for electricity, heating and transport.

The inefficiency of converting electricity into hydrogen then back to electricity is what fundamentally disqualifies hydrogen from being the cheapest option as an energy carrier here and now in the 2020's. That being said, this statement only stands true if practical matters don't make that hydrogen cheap despite its inefficiency. Namely: could electrolyzers be *paid* by an electrical grid to mop up the spikes in generation inherent to wind- or solar-powered renewables. Inefficiency alone never stopped fossil fuels from dominating our energy use.

In our discourse here we did not by any means attempt a comprehensive review of the topic, but merely introduced its broad strokes to help appreciate the pivotal role pipelines must fulfil. There are, for example, people who insist that storing ammonia or methanol might make better sense than hydrogen. There's hydrogen for aviation. Doubtless there is much more as well.

# E4 – WHO TRANSPORTS HYDROGEN?

Government whitepapers advocating a hydrogen economy point to a chicken-and-egg situation. To bring the picture into focus let's critically examine the evolving networks connecting available suppliers with their users' demands.

Unlike oil and gas, today's hydrogen production isn't traded on an open market. There isn't yet a mechanism for labelling its origin, nor how clean it is. Slowly but surely, initiatives and standards are emerging to track a *"hydrogen parcel"* and certify the extent of its clean credentials – one recently in the news does so via blockchain (because of course it does). It is manufactured, moved and consumed piecemeal, treated as an *"industrial gas"*. To make it a local, regional or global market commodity, stakeholders are having to collaborate.

It is seen as likely, given the abundance of renewable energy in parts of the world far removed from the centers of population and industry, that hydrogen will be imported across continents, much like liquefied natural gas (LNG). For this reason, countries that are most entrenched in an oil-based economy, like Saudi Arabia, are those most interested in pursuing hydrogen production for export.

It's already here today. Hydrogen is being blended into natural gas networks in the UK. Pure hydrogen transmission and distribution infrastructure is being supplemented by new pipelines, converted or newly built all over the world. It would be remiss of any pipeliner to ignore it. That includes liquid pipeliners, as there are situations where hydrogen might be pipelined as a liquid batch, bound within a *liquid organic hydrogen carrier* (LOHC) or in the form of methanol or ammonia. Frankly, discussing hydrogen transport with adequate figures warrants its own book and maybe a PhD.

Transporting hydrogen using pipelines is a topic we shall visit more closely in a later section of this appendix, examining their technical implications in much more detail.

# E5 – Equitable economics

It would be an absolute travesty if any transition to a greener world served to exacerbate the divides between rich and poor. A lost generation is emerging; it's not just global-scale developed, developing and lagging economies. Even in the same country, deep-seated resentment is boiling over. Household finances already don't stack up as it stands. At time of writing, the UK places a heavier tax burden on home insulation than gas for the boiler! If one rents, leases or buys an electric car, they are not the poorest in society, yet they would receive substantial tax relief. Our tax policies, as they stand, are regressive. We should take every opportunity to make that better, always remaining aware of how easy it is to inadvertently make matters worse.

Who pays to build the infrastructure for greener electricity and hydrogen? It's unfair to spread costs evenly if the benefits are reaped chiefly by the wealthiest. Whitepapers on the subject focus on the implications of government policy. A transition of this kind can only happen sensibly with proper regulations in place.

# E6 – The state of the art

Hydrogen can be pipelined in large quantities and held in long-term storage facilities. Hydrogen pipelines have existed for decades. Around 1 000 miles of hydrogen pipelines operate in Europe already, and a similar length operate in North America. These link facilities where hydrogen is manufactured – mainly from natural gas today – to its consumers: heavy chemical industries and refineries. The bigger asset is pre-existing natural gas pipelines and storage.

In some places, pipelines have been converted outright from natural gas to pure hydrogen service, not for heating but for industry. In Germany, this was done for a pipeline linking the Lingen power plant (collocated with a link from offshore wind turbines) to major consumers of hydrogen: refineries in the Ruhr area. This is estimated to avoid 100 thousand tons of $CO_2$ emissions.

In other places, countries see hydrogen as part of an even wider transition beyond natural gas. The UK is adapting pipelines across its transmission and distribution networks to accommodate blended hydrogen in large proportions.

Hydrogen poses technical challenges, but these are already well known: making and using hydrogen is nothing new. Being the smallest element, an $H_2$ molecule consists only of a pair of protons and a pair of electrons. This means materials that contain most other gases are permeable to hydrogen: special materials are essential for handling it.

Special measures are also needed to guard against *embrittlement*, so pipeliners do consider whether the lifespan of their materials might be affected. But hydrogen pipelines are not the barrier.

Hydrogen's biggest hurdle is cost of manufacture. Scaling up essentially rules out small electrolyzers at every point of use. A hydrogen economy deliverable today would hinge on generating it at large-scale, centralized facilities. That is the first phase of a hydrogen economy, and it is the only phase approaching being realized today.

# E7 – The Proposed Transitions

Proponents say that *"any hydrogen is better than no hydrogen"*, claiming that blue hydrogen could serve as a transition technology, facilitating the construction of useful infrastructure to get past the chicken-and-egg. Pipelines, storage, and users can then pave the way to generate, distribute and consume *greener* hydrogen.

Detractors retort that building infrastructure in such a way, relying on vast supplies of hydrogen coming from centralized facilities, might only cause us to prolong our dependence on fossil fuels, remaining hooked on grey hydrogen.

By examining a hydrogen distribution network, a pipeline simulator helps settle the dispute between proponents and detractors of a hydrogen economy. It substantiates proposed ways forward for an existing or planned grid, and how it evolves over time to accommodate new inputs and offtakes. If a specific pipeline is built around a centralized generation facility, will it be amenable to becoming flexible enough for distributed green hydrogen to one day take over?

Large-scale adoption of hydrogen involves changes by users and a nationwide transmission and distribution system. The economics of electrolyzers are a barrier, particularly as they won't run full-time, but challenges like that may be overcome.

> *"Simply plugging massive numbers of wind farms into the existing [power] grid and hoping for the best is unlikely to be a successful approach."* [3]

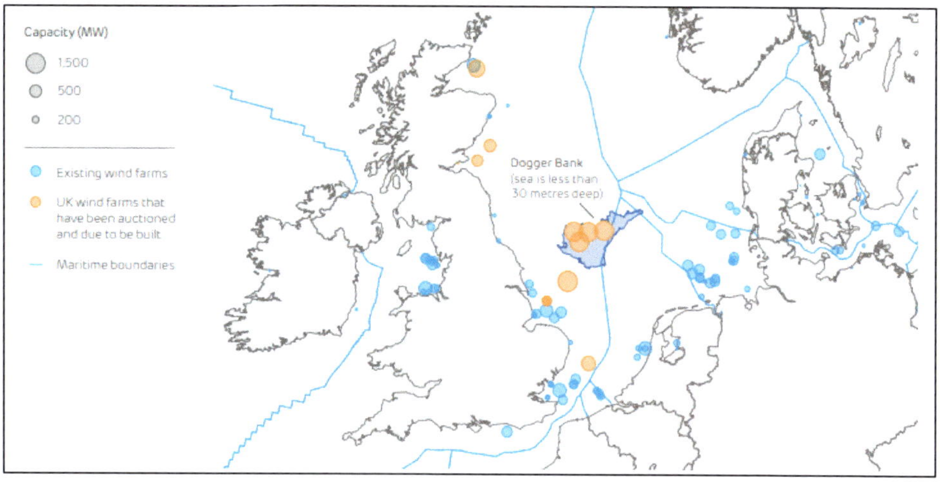

Figure 108 – Wind farms in the North Sea are moving further offshore (electricinsights.co.uk)

The UK's anticipated excess electricity production in 2030 could be used to make 670 thousand tons of hydrogen, 1% of the total global industrial consumption at time of writing (2020), enough to heat 2 million homes. As wind farms move further offshore, long-distance high-voltage grid connections make up a third of total cost. Electrolyzers could make hydrogen directly out at sea, using electricity at offshore wind turbines to split water. Hydrogen could be pipelined to shore using existing oil and gas pipelines, which are expected to have spare capacity as North Sea gas fields are winding down. [3]

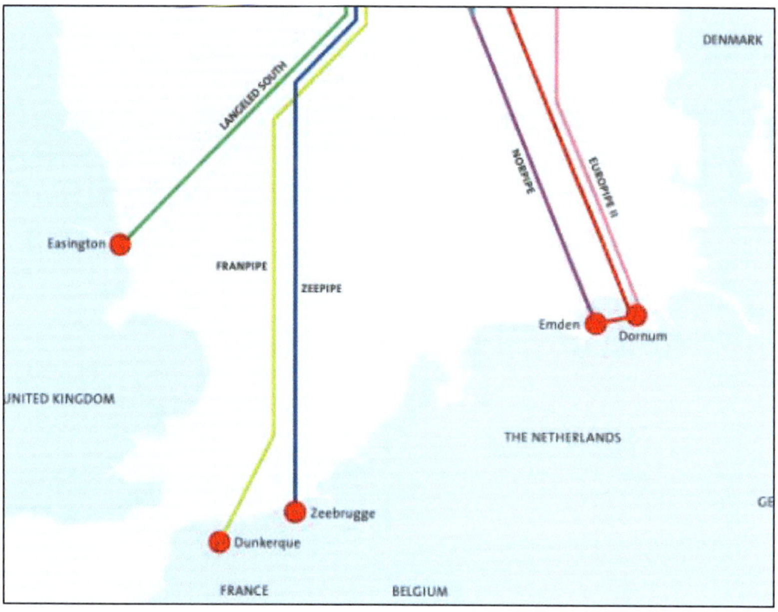

Figure 109 – Oil and gas pipelines under the North Sea (Image Credit: Gassco, Norway)

# E8 – THE TECHNICAL DETAILS

## Gigawatt speed limits: flow in a hydrogen pipeline

> *"Due to the high calorific value and compressibility of hydrogen, an extraordinarily high energy density can be achieved. In comparison to a 380 kV double system overhead line with 1.5 GW, a gas line (PN 80, DN 1000) can transmit up to ten times the power in natural gas and hydrogen operation at a fourteenth of the specific costs."* – Siemens Whitepaper, 2020

If hydrogen is not produced directly at point of use, it must be transported. To verify high-level claims like the quote above, that pipelining hydrogen is so much easier than transmitting electrical power, pipeliners must consider many factors. Those factors include the material every valve and fitting is made of, the throughput of the pipeline itself, and compressor thermodynamics. The latter two avenues of enquiry are aided by pipeline simulators such as Atmos SIM.

It is good to be skeptical, believing arguments put forward by vested interests only once we see for ourselves how they play out in our simulator. If a claim turns out to be false, it's not indicative of malice, and we gain insight! Wouldn't reinforcing trunk transmission pipelines and local distribution networks be a very costly endeavor? The cost of uprating every pipe in every street, service pipes to every property and possibly every pipe inside every property. Previous mains and service pipe renewal had inserted smaller plastic pipes *inside* the existing old cast and ductile iron. So today, much of the UK's low-pressure network has *less* than original capacity! Won't that need upgrading?

There is a common perception that hydrogen is a tough fluid to pipeline. A rule of thumb says it can only carry a third of the energy through the same line, so a pipeline's throughput would need trebling – no mean feat. However, that is not strictly the case. As more hydrogen is blended into a natural gas pipeline, its low density does reduce throughput capacity, but not in such a drastic way. We must look more closely before believing that simple rule of thumb. *"For every complex problem there is an answer that is clear, simple, and wrong."*

Hydrogen is the lightest element, and this means that by standard volume it contains only a third (3.1 kWh/Sm$^3$) of the calorific value of methane. But its calorific value by mass (33 kWh/kg) is more than double that of methane. Neither of these values alone determines a pipeline's maximum throughput flow. Assuming the compressors are coping, can the same pipeline transmit enough hydrogen at an energy per *actual volume* (kWh/m$^3$) comparable to that of natural gas?

At lower density, frictional losses increase *more than linearly*. The significance of this statement is that a light gas must be pressurized disproportionately more intensely at an inlet, so that it traverses the pipeline at the same mass flow rate. Add to that the inefficiency of any equipment compressing a light gas, and we'd expect to draw more power transmitting hydrogen than natural gas.

Would hydrogen reduce a pipeline's net energy flow rate? What maximum velocity and what pressures are needed?

Using Atmos SIM let's compare a flat 100 km pipeline in a few scenarios, pausing to note that the energy carried by this simple 1-metre diameter gas pipeline is equivalent to a quarter of the UK's entire electrical generation capacity at time of writing. In each column, we state the net calorific value as an energy flow rate, in units of gigawatts. 10 GW can power a billion lightbulbs.

| **Net energy flow GW Outlet velocity m/s ↘** | **Original ΔP 70 to 50 barg** | **Twice the ΔP 70 to 30 barg** | **Hi-P, same ΔP 170 to 150 barg** |
|---|---|---|---|
| **Methane** | 18 GW 12 m/s | 24 GW 26 m/s! | 32 GW 6 m/s |
| **Hydrogen** | 14 GW 36 m/s! | 18 GW 77 m/s! | 23 GW 20 m/s |

**Table 9 – Energy flows and velocities of methane and hydrogen at various pressure drops.**

We see the speed limit for gas flow velocity is being violated by pure hydrogen!

Perhaps pressure drop might not be the controlling factor after all, but rather our attention should focus on maximum velocity? This would be an unfamiliar position to find ourselves in: faster velocities involve more frictional losses and substantial pressure drops, which goes against optimizing initial upfront costs versus ongoing operating costs. Usually to design a long distance pipeline we expect pressure drop to be of utmost importance. But let's run with it and see.

A typical rule of thumb is that natural gas pipelines should not exceed 18 m/s, and a working assumption is often 15 m/s. But this is a rule of thumb, not a universal law of nature. It doesn't apply to hydrogen. Pipelines carrying saturated steam, for instance, routinely operate between 25 and 40 m/s. [106] So what is a typical practical speed limit along a long-distance hydrogen pipeline? What else is limiting a hydrogen pipeline?

Perhaps due to their rarity, a rule of thumb for flow velocity in a hydrogen pipeline is very hard to come by.

From the perspective of flow-induced turbulence and flow-induced pulsations, faster velocities are possible with hydrogen. However, intrusive elements such as thermowells or certain types of flow meter face a worse risk of failure when exposed to faster flow velocities, even at similar levels of flow kinetic energy. [107]

There are practical considerations. Most manufacturers now state that the current turbine meters are suitable for a maximum of 10 vol% blend of hydrogen. For a turbine flow meter, a typical product specification in 2020 is a maximum actual flow rate of 24 000 $m^3$/h for the largest available bore diameter (600 mm). This limit applies to all gases, regardless of composition. Consequently, a turbine meter imposes a fixed maximum velocity on a pipeline of 24 m/s. Beyond that, mechanical parts of the turbine meter may become damaged, or erosion could degrade the surface of its rotating wheel. If this limit is violated, replacing natural gas with hydrogen at the same energy flow rate may require a rethink of the measuring principle in a replacement flow meter. Manufacturers, of course, are conducting research to handle faster velocities.

Turbine meter manufacturers also specify a *minimum* flow rate, which is made more stringent (i.e. higher) the more hydrogen is present. This narrows the measurable range of flow velocities as hydrogen is blended into natural gas.

In all cases, it seems reasonable to accept a natural gas pipeline that was flowing at a velocity of 12 m/s can become a hydrogen pipeline flowing at 20 m/s. But if we attempt to operate at the original pressures, hydrogen would flow at a velocity of 36 m/s, which might destroy a typical flow meter! As demonstrated by the simulated initial scenario (methane 18 GW 12 m/s) and the final scenario (hydrogen 23 GW 20 m/s), the focus for gas pipeliners looking at blending or converting to hydrogen is how to operate at higher pressures throughout the line, rather than handling larger pressure drops. A higher-pressure delivery is particularly useful as hydrogen is meant for equipment at the customers site, and that equipment demands hydrogen must be at higher pressure than natural gas.

# Frictional losses along a hydrogen pipeline

Since gases generally flow in the far-turbulent regime, frictional losses are minimally if at all affected by viscosity. The viscosity of hydrogen is lower than that of natural gas. Blending more hydrogen into a natural gas mixture reduces its viscosity, and therefore there are lower frictional losses for a given actual volumetric flow rate ($Q$). However, to transmit the same amount of energy through a pipeline, more actual volume must flow through, which will amount to a higher flow velocity. The overall net effect is that higher frictional losses are incurred along the same pipeline if it is transporting the same energy in the form of a higher hydrogen gas blend. [108]

# Powering compressors on a hydrogen pipeline

Sometimes, the limiting factor preventing a pipeline from operating at higher pressures throughout its entire length is not *maximum allowable operating pressure* (MAOP), but the power rating of a compression station. Compressors placed along a well-run high-pressure pipeline will often already operate near their *best operating point* (BOP). Local medium- and low-pressure distribution grids rely on regulating pressures *down* without any compressors of their own.

A reciprocating compressor is the most efficient way to pressurize pure hydrogen, [109] but it is inherently designed to handle a narrow range of flow rates, rendering it unsuitable for a pipeline in which the proportion of hydrogen fluctuates over time. A centrifugal compressor can achieve the same pressure ratio compressing any proportion of hydrogen blended in with natural gas, up to and including pure hydrogen. But to compress pure hydrogen it must operate at a rotational speed 1.7 times faster. Unfortunately, such speeds are limited by a material's strength. [110]

In addition to having to operate at higher speeds, a centrifugal compressor must also be capable of providing enough power to achieve the same pressure ratio ($P_2/P_1$) with a hydrogen + natural gas blend, because hydrogen is inherently more difficult to compress than natural gas. The theoretical minimum compression work needed to pressurize any gas in an ideal, perfectly efficient ($\eta_s = 100\%$) equipment is a function of the *heat capacity* ($c_P$) of that gas, and its *ratio of heat capacities* ($\gamma = c_P/c_V$). Work is expressed as *imparted head* ($\Delta H_{\mathrm{J/kg}}$):

$$\frac{P_2}{P_1} = \left(1 + \frac{\eta_s}{c_P \cdot T_1} \cdot \Delta H_{\mathrm{J/kg}}\right)^{\frac{\gamma}{\gamma - 1}}$$

We see that it is also a function of the inlet temperature ($T_1$), which we shall discuss shortly, but the dominant effects are the higher heat capacity of hydrogen compared to natural gas ($c_P = 14.3$ vs. $2.3$) as well as the higher heat capacity ratio of hydrogen ($\gamma = 1.4$ vs. $1.3$). Combined, this means that a given amount of work done on a hydrogen blend achieves a lower pressure ratio than would be seen in natural gas. To compress a gas containing more hydrogen to a given pressure ratio, more work must be put in to a compressor. And all of this is before considering any inefficiencies. [107]

Therefore, for a pipeline to transmit the same energy in the form of a higher (or lower) hydrogen + natural gas blend, its compression stations must be capable of accommodating new *operating points* that impart more power to a higher flow. This must be planned for very carefully to ensure every compressor in a station can ramp up and down to deal with fluctuations in gas composition while maintaining an acceptable efficiency. [107]

Every pipeline is unique, but some broad conclusions can be drawn. A typical gas pipeline was simulated in a paper presented at the Pipeline Simulation Interest Group 2021 conference, and it was found that a 10% blend of hydrogen consumes 25% more power in its compressors, and a 20% blend of hydrogen consumes 70% more power than the equivalent base case scenario (transporting the same calorific value by setting energy flow as the constraint). These numbers will vary depending on the spacing and pressure ratios of compressor stations in a given pipeline. Pipelining a higher hydrogen blend always consumes more compression power. [107]

For this reason, it has been proposed that if a pipeline is to transport hydrogen produced by electrolysis as a way to make use of surplus electricity during peak periods, an approach called power-to-gas (P2G), it might be advisable to use electrically driven compressors rather than drivers that rely on combustion of part of the pipeline gas (turbines or engines). This would preserve the hydrogen blend for uses that are arguably in more urgent need for it, and offer further capability to draw electrical power at times when the grid is desperately trying to get rid of it. This approach is referred to as power-to-pressure (P2P). Expanding hydrogen generation, storage and transport raises a fair number of questions whose studies require simulation. [111]

This is part of a broader need for coordinating the gas and electric markets: they're more coupled than ever. Many parameters can only be simulated and optimized by using dynamic models of gas pipelines and electrical power grids simultaneously. [112]

It is particularly important as many countries are planning to keep increasing the amount of renewables available to dispatch as a priority, leaving pipeline capacity that had been utilized to fuel gas-fired thermal plants available. The number of variables at play in these scenarios is unprecedented, and they are absolutely crucial to selecting the right compression technology even before considering the change of gas composition to blend in more hydrogen. [113]

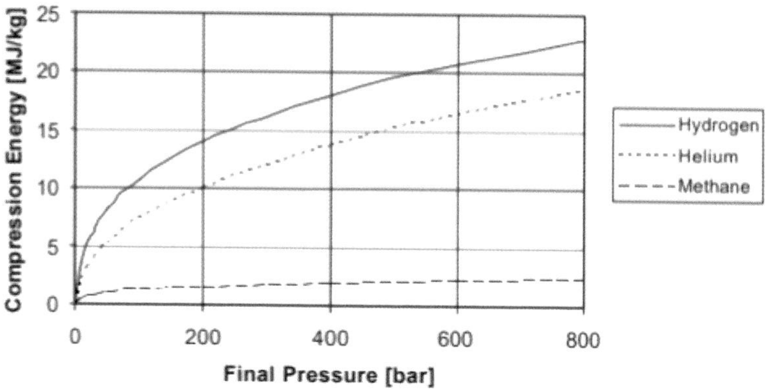


Figure 110 – A theoretical perfect compressor (isentropic, i.e. reversible adiabatic) must do a lot more work to compress hydrogen than it does to compress the same mass of methane. This is partly made up for by hydrogen's better combustion value, but it does remain a challenge. [113]

The exact calculation of compression power is quite complex, as we will discuss in a chapter of the main book (*"Pumps and compressors"*). However, an indicative figure for pure hydrogen is that compressing it from atmospheric pressure to 200 bar consumes 7% of its combustion value, based on a manufacturer's stated figure for 1 000 kg/h. This is approximately in line with what we would expect using the ideal gas law, as it lies between the isothermal and isentropic (reversible adiabatic) figures. [113]

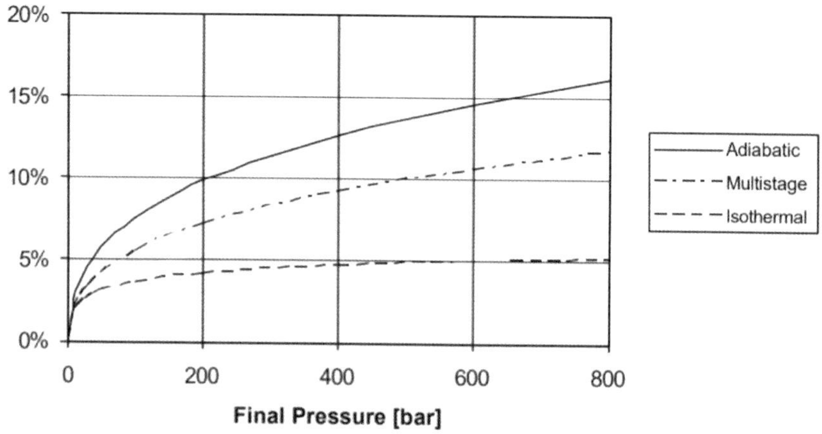


Figure 111 – The proportion of hydrogen's combustion value consumed in compressing it, assuming that a pipeline compression station uses the pipeline fluid to fuel compressors. [114]

# Temperature profile along a hydrogen pipeline

The Joule-Thomson effect causes the temperature of any gas to change as it flows along a pipeline, because there is always a pressure gradient driving the flow, so the density drops. Hydrogen is unique in that the Joule-Thomson effect does not cause it to cool down as its pressure drops the way natural gas does, but instead it causes it to warm up. And conversely, an ideal perfectly efficient compressor would cause pure hydrogen to cool down rather than to heat up.

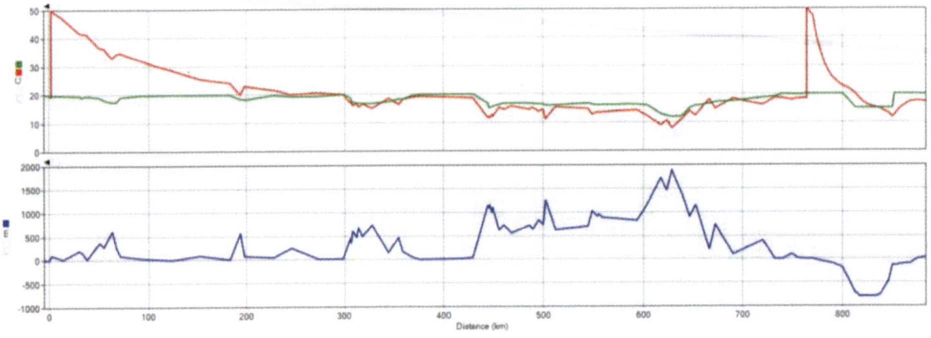

**Figure 112 – The Joule-Thomson effect causes a natural gas pipeline to operate cooler than ambient temperature. The same effect causes flowing hydrogen to be warmer than ambient.**

# Linepack within a hydrogen pipeline

Linepack is gas compressed within a pipeline as short-term buffer storage. This allows fluctuating supplies to be balanced against fluctuating demand patterns. By exploiting linepack, injections and offtakes can, within limits, be changed at will independently of one another. This couples installed capacity with available linepack as an optimization task we explore in the final chapter of the main book.

The available linepack of a section of pipeline depends on its pressure, which in turn depends on the flow rate. Therefore, although maximum and lowest allowable linepack are fixed constants, the available linepack varies over time.

If the maximum throughput of hydrogen flows at steady state through the entire 1 000 km length of a 900 mm diameter pipeline, with a completely flat elevation profile, going from 100 to 40 barg, then the available linepack amounts to 18 million $Nm^3$ of hydrogen, storing the equivalent of 65 GWh.

# So should we build it, and if so, how?

The natural gas grid is already coupled to electrical power cables in the many places that rely on gas-fired peaking power plants: turbines and engines, large and small. The emerging potential of hydrogen as an energy carrier represents the first time that a gas pipeline system is coupled to an electrical power grid in the other direction – by electrolyzers whose purpose is to generate hydrogen only during periods of surplus power. This introduces a typical *"supplier pattern"* to a gas pipeline every time it is fed by hydrogen from an electrolyzer.

---

What if the hydrogen were injected at a different station?

Whether it is a good idea to even install a facility at a location is not an easy decision to make. Wanting hydrogen to reach a customer according to a certain demand pattern doesn't tell us if it is best to install an electrolyzer here, a gasifier there, or a reformer elsewhere. And how much generating capacity should that installation be sized for? The answer depends on the pipeline's linepack, as well as decisions on whether storage facilities are built, and where.

In the chapter on optimizers, in the main book, we look at meeting this challenge. We see a complex optimization problem being posed, and that is where computers excel; how do we consider the best way forward for a hydrogen pipeline, be that in a pure form or blended with natural gas? It seems that there would be value in modelling electrical power grids and pipelines within the same simulator, especially if down the line hydrogen does end up becoming such a major energy carrier that a two-way coupling happens where it is fed by electrolyzers according to supplier-patterns and generates electricity for a power grid according to demand-patterns. This may seem outlandish to many today, yet even if their view is right, we gain insight by seeing exactly *how* an idea is bad.

Nonetheless, we should not be too keen to branch off in that direction and lose sight of the strength of a good pipeline simulator in the first place: its focus.

# Simulating a hydrogen pipeline

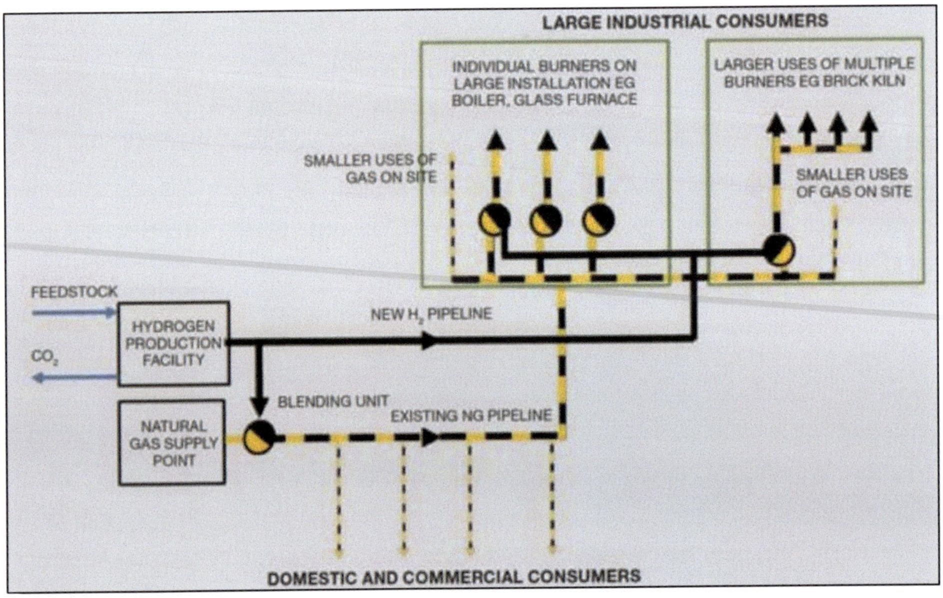


**Figure 113 – Proposed pipeline infrastructure for blended natural gas and pure hydrogen (Credit: the HYNET carbon capture and storage project in North West England)**

The industry is still waiting for a sizeable natural gas pipeline to start carrying a significant proportion of hydrogen (5% or more) so everyone can definitively validate our results against what happens in reality. One of the best contenders is the HYNET project in North-West England, but although they have identified the broad strokes of that pipeline's start and end points, at time of writing the actual route selection and detailed studies are not yet finalized.

A typical pipeline presented as a case study and which we drew upon in our discussion in the preceding sections is available at the following reference. [107]

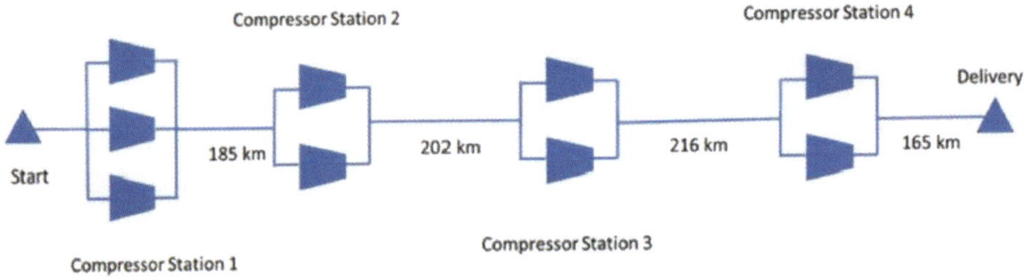


**Figure 114 – A typical natural gas pipeline selected for a hydrogen blending case study.** [107]

# APPENDIX F – QUIZ

1. Did the Romans use valves?
    a. Valves were rarely used. In open channels flow can only be diverted.
2. Would you be able to pick up a barrel of crude oil?
    a. Depending on density, it would weigh 110 to 150 kg.
3. Which weighs more: a pint of crude oil, gasoline, or milk?
    a. Milk is the heaviest, then crude oil, then gasoline is the lightest.
4. What do the units of MMSCFD refer to?
    a. Million standard cubic feet per day – the standard flow rate of a fluid.
5. What are the units of pressure in the SI and kilogram–meter–second systems?
    a. $1\ \text{Pascal} = 1\ \text{kg} \cdot \text{m}^{-1} \cdot \text{s}^{-2}$
6. What three physical laws of conservation underlie every pipeline simulator?
    a. Conservation of mass, momentum and energy.
7. An equation of state relates pressure and temperature to what fluid property?
    a. Density
8. Any gas abides by the ideal gas law as its equation of state in what conditions?
    a. Near-atmospheric pressure and near-room temperature
9. At the end of a pipeline, pressure is lower, so the density of a gas is lower. What does this imply about velocity along a gas pipeline of constant diameter?
    a. The flow velocity becomes progressively faster along the pipeline
10. Can a steady model capture kinetics?
    a. Yes: it just assumes that heat transfer (for instance) is in equilibrium.
11. Can many steady states in succession capture evolving dynamics?
    a. Successive steady states cannot do so. A transient model is needed.
12. How can we validate a simulation is capturing what really happens?
    a. Validation checks such as simulating again with a tighter mesh (knots and steps), comparing against site data at a meter the model does not know about, and certain analytical solutions.
13. How accurate can we expect the physics of an offline scenario to be?
    a. Usually as accurate as the underlying equations – such as the friction factor correlations – and physical model assumptions.
14. How accurate can we expect meters and state estimators to be online?
    a. It depends on meters and instrumentation – typically 0.2 bar.
15. How can we keep an online simulator close to reality during operation?

    a. By continuous learning of model parameters that vary, like the flow meter offsets, using automatic online tuning mechanisms.

16. Can an *"incompressible mode"* system live on site deduce pressures?
    a. Yes, but won't reconcile flow rates: it just follows a fiscal meter.
17. What carries more energy along a pipeline of the same diameter at the typical limit of pressure boundary conditions, crude oil or natural gas?
    a. Through 100 km 1-metre pipeline from 70 to 50 barg, methane flows at the equivalent of 18 GW, light crude flows at 45 GW.
18. Can the direction of gas flow in a pipeline be against pressure drop?
    a. Yes, gas is dense at high pressure, so if it is flowing downhill it might be flowing against pressure drop, just like in a liquid line.
19. Is there any pipeline in which the viscosity of the fluid is not very influential?
    a. Yes, in a gas pipeline, as Reynolds number is typically very large
20. What is the ratio between the dynamic viscosity and kinematic viscosity?
    a. Density
21. Do suppliers prefer to supply at a fluctuating flow rate or at a constant rate?
    a. It is typically more economical for them to supply at a constant rate
22. If two streams of differing composition are blended, is their combined actual, standard or mass flow rate going to be a simple sum of the two?
    a. For most pipeline fluids, yes, except if there is non-ideal mixing, as is the case when blending ethane with propane for example.
23. Why is it better to adjust a liquid pipe's efficiency than its wall roughness?
    a. Because in some flow regimes, wall roughness has little to no effect.
24. Does inventory affect transit time for the same standard flow of gas?
    a. Yes, at the same standard flow rate, the transit time through a pipeline is directly proportional to the inventory of gas it is holding as linepack.
25. Can a pig move faster than the fluid flowing in the pipeline?
    a. A pig can move faster than gas along a downhill slope, as the solid is denser than gas. In a liquid line it is less likely to do so.
26. Which land animal's heart is largest, and which has the thickest walls?
    a. An elephant heart pumps the greatest volume of fluid, so it is the largest. A giraffe's heart has to lift blood to the highest levels, so it has the thickest walls (3 inches thick, achieving 0.3 atmospheres).
27. Does a perfectly efficient isentropic compressor affect temperature?
    a. Density changes in any compression, affecting temperature.
28. Does an incompressible liquid in an ideal pump change temperature?
    a. No, the density does not change, and there's no inefficiency.

29. Do reciprocating pumps present a net positive suction head required?
    a. Yes, their moving parts might bring about excessive cavitation.
30. Do all pumps and compressors require recycling below a certain speed?
    a. No, reciprocating pumps and compressors don't have a surge limit, though most pumps do have a net positive suction head required.
31. Are reciprocating compressors ever preferable to centrifugal compressors?
    a. Yes, in applications with low flow rates and high compression ratio
32. On a centrifugal pump or compressor map, what kind of flow rate is general?
    a. Actual volumetric flow rate (actual flow)
33. Does gas composition affect the performance of a centrifugal compressor?
    a. Yes, because it affects the density and actual volumetric flow rate
34. What's the effect of placing centrifugal pumps or compressors in series?
    a. The overall head-flow curve of a series combination is flatter.
35. What's the effect of placing centrifugal pumps or compressors in parallel?
    a. The overall head-flow curve of a parallel bank is steeper.
36. For smallish changes what do the Navier-Stokes equations simplify to?
    a. A wave equation.
37. Does the gas pressure at a delivery point remain approximately the same?
    a. In a gas pipeline, it will typically fluctuate significantly as the linepack accumulates and unpacks over time. Regulator valves reduce a delivery pressure to the level required by a consumer.
38. What happens to flow when pressure at the outlet of a pipeline suddenly falls?
    a. The pressure drop across the pipeline increases so the flow increases
39. Do frozen look-aheads hold everything constant at controlling items?
    a. No, they hold flow rates constant, allowing pressures to vary.
40. Is an optimizer only able to examine steady states?
    a. An optimizer can also be built to simulate transient scenarios, analyze their results, improve a next guess and simulate again.
41. Is a liquid more volatile if it has a lower or a higher vapor pressure?
    a. A higher vapor pressure means more of the fluid is present in its vapor phase at a given pressure – i.e. a more volatile liquid.
42. Does the vapor pressure vary in an incompressible liquid?
    a. Yes. Its density is taken to be constant at all pressures, but the temperature changes, which strongly affects vapor pressure.
43. Is a slack model capable of multi-phase flow?
    a. Slack is when part of a homogenous fluid is in its vapor phase. Multi-phase flow could refer to other very different phenomena.

# INDEX

# REFERENCES


[1] The Economics of Nuclear Energy, Real Engineering on Youtube
[2] Breakthrough in renewable energy - VPRO documentary
[3] Electric Insights Report, November 2020, Drax
[4] Greenhouse Gas Emissions Study of Australian CSG to LNG, WorleyParsons, 2011
[5] Ancient Inventions, Peter James, Nick Thorpe, Ballantine Books, 1995
[6] itia.ntua.gr/ahw/work/73/
[7] Vitruvius, The Ten Books on Architecture, 8.6 (translated by Hicky Morgan)
[8] Kurlansky, M. (2003). Salt: A World History. Penguin Books
[9] Scientific American Vol 268 No 6, Hans Ulrich Vogel, 1993
[10] https://bakucity.preslib.az/en/page/HVoDG2TkaP
[11] https://csegrecorder.com/articles/view/ancient-chinese-drilling
[12] The salt industry of Tzeliutsing, Wallace Crawford, 1926
[13] Islamic Technology: An Illustrated History, Ahmad Y. Al Hassan and Donald R. Hill. 1991. Cambridge
[14] Structures or why things don't fall down, J. E. Gordon
[15] "A city shaped by steam", New York Times, 2014
[16] www.historywebsite.co.uk/articles/Gas/Distribution.htm
[17] https://www.grc.nasa.gov/www/BGH/entropy.html
[18] minenergy.gov.az/en/neft/neft-senayesinin-inkisaf-tarixi
[19] visions.az/en/news/618/3b2f9122/
[20] History of the Darcy-Weisbach equation for pipe flow resistance, Glenn Brown, Oklahoma State Univ.
[21] slb.com/resource-library/article/valve-insights/how-it-works-actuators
[22] History of Hydraulics, by Hunter Rouse and Simon Ince. Iowa Institute of Hydraulic Research, 1957
[23] Less is More: Accuracy vs Precision in Modelling, Susan Bachman and Mary Goodreau, PSIG 2000
[24] Dynamics of Physical Systems, Robert Cannon, 2012
[25] Pipeline Rules of Thumb Handbook, McAllister, 2014
[26] Unified Friction Formulation from Laminar to Fully Rough Turbulent Flow, Applied Sciences 2018, Dejan Brkić and Pavel Praks
[27] Advanced Natural Gas Engineering, Economides, 2013
[28] https://www.aboutpipelines.com/en/pipeline-101/why-spills-happen/
[29] What Went Wrong? Trevor Kletz, Elsevier, 1998
[30] Single-phase flow assurance, Ove Bratland, 2013 drbratland.com
[31] thechemicalengineer.com/features/clean-hydrogen-part-2/
[32] Petition for diluted bitumen rulemaking – EPA (accessed online on 24 Aug 2021)
[33] Hazardous Liquid Pipelines - Basics and Issues, Pipeline Safety Trust, 2015
[34] Review of Vapor Cloud Explosion Incidents, UK HSE, 2017
[35] McKay F.F. et al., Hydrocarbon Processing, November 1977, 56(11):487
[36] www.ipcc.ch/site/assets/uploads/2018/02/ipcc_wg3_ar5_chapter7.pdf
[37] Net Zero and Safety, Gilmour, The Chemical Engineer, Dec 2019 / Jan 2020
[38] US is Overflowing with Natural Gas, not Everyone can get it, WSJ, 2019
[39] Natural Gas – Fuel for the 21st Century, Smil, 2015
[40] Oil and gas sector can bring quick climate win by tackling methane emissions, UNEP, 2019
[41] pgjonline.com/news/2020/12-december/eia-natural-gas-venting-flaring-set-records-in-2019
[42] Energy: A Human History, Rhodes, 2018.
[43] Equation of State for Natural Gas Systems (PhD Thesis), Uribe, 1999.
[44] https://tsapps.nist.gov/publication/get_pdf.cfm?pub_id=903432

[45] reutersevents.com/renewables/solar/understanding-energy-density-thermal-storage
[46] Mass and Heat Transfer, Russell, 2019
[47] Crane Technical Paper 410: Flow of Fluids Through Valves, Pipes and Fittings, 1999
[48] SmarterEveryDay on YouTube
[49] Pipeline Leak Detection Handbook, by Morgan E. Henrie, Philip Carpenter, and R. Edward Nicholas
[50] 1984 PSIG talk by Modisette, Nicholas, & Whaley
[51] Preissmann, A. (1961) Propagation des intumescences dans les canaux et rivières. First Congress of the French Association for Computation, Grenoble, 433-442
[52] On Simulation Accuracy, Per Lagoni and Jon Barley, PSIG 0711
[53] Pipeline Variable Uncertainties and Their Effects on Leak Detectability, API 1149, 2nd Ed. (2015)
[54] Pipeline System Automation and Control, Mike Yoon, C. Bruce Warren and Steve Adam, ASME 2007.
[55] Simulation of Rapid Transients in Gas Pipelines for ESD Valve Design, Jason Modisette, Garry Hanmer, PSIG 2021
[56] Aalto M., K. I. Keskinen, J. Aittamaa, and S. Liukkonen, Fluid Phase Equil., 114: 1 (1996)
[57] Tullis, J. Paul. Hydraulics of Pipelines: Pumps, Valves, Cavitation, Transients. United Kingdom: Wiley, 1989.
[58] Managing Liquid Pipeline Slack, Pipeliners Podcast, Jan 2019
[59] An Investigation into how to Optimize the Effect of DRA on the Flow rate & Pressure of Single Phase Fluid Pipelines, Naffaa, PSIG 2019
[60] popularmechanics.com/science/energy/a7480
[61] Design calculations for oil and gas pipelines, Kasch, 2007
[62] linkedin.com/pulse/cavitation-misunderstandings-other-stories-simon-bradshaw
[63] Chemical Engineers Handbook, Robert Perry and Cecil Chilton, 5.ed 1973
[64] Controlling Positive Displacement Pumps, Walter Driedger, 2000, accessed via pumpfundamentals.com
[65] Adapted from a video by WR Training: youtu.be/UmoOXdLLITQ
[66] Hydraulic Design for Business Development Professionals, Pipeliners Podcast, Apr. 2019
[67] Coulson and Richardson's Chemical Engineering Volume 6 (Design), 2ed, by Ray Sinnott, Pergamon Press, 1994
[68] Process Centrifugal Compressors, Lüdtke, 2004
[69] A Working Guide to Process Equipment, Liebermann, 2014
[70] Handbook of Natural Gas Transmission and Processing, Mokhatab 2019
[71] Numerical Simulation of Gas Pipeline Networks, Il'kaev, Seleznev, Aleshin, Klishin, 2005
[72] Fundamentals of Pipeline Engineering, Vincent-Genod, 1984
[73] Jet Propulsion, Cumpsty, 2015
[74] Operation of Centrifugal Compressors in Choke Conditions, Kurz et al, Turbomachinery Symp. 2011
[75] Pipeline Design & Construction, Mohitpour, 2000
[76] Natural Gas Processing, Bahadori, 2019
[77] Oil and Gas Pipeline Fundamentals, John L. Kennedy, 1984
[78] Oil and Gas Pipelines in Non-Technical Language, Thomas Miesner and William Leffler, 2nd ed. 2020 PennWell
[79] thetyee.ca/Opinion/2018/11/28/Air-Barrels-Pipeline-System/
[80] *Shock to the System*, US Chemical Safety and Hazard Investigation Board, youtu.be/_icf-5uoZbc
[81] Modelling of Rapid Transients in Gas Pipelines, Hanmer, Mora, de Marco & Lacerda
[82] ntsb.gov/investigations/AccidentReports/Reports/PAR0202.pdf
[83] tandfonline.com/doi/abs/10.1080/1573062X.2020.1729388 Preliminary sizing of surge vessels on pumping mains, Cogent engineering, Nuralhuda Aladdin Jasim, 2020
[84] Adding it Up, Barrett, The Chemical Engineer, Feb. 2019
[85] Simple Method Predicts Gas-line Blowdown Times, Weiss, Botros, Jungowski (NOVA) – O&GJ, 1988
[86] Bhopal Leaves a Lasting Legacy, Kletz, chemicalprocessing.com/articles/2009/238/
[87] https://www.aboutpipelines.com/en/pipeline-101/why-spills-happen/
[88] Protection of piping systems subject to fires and explosions, UK HSE, 2005
[89] pipelinerspodcast.com/episode-27-rupture-detection-giancarlo-milano/
[90] https://www.ntsb.gov/investigations/AccidentReports/Reports/PAR0202.pdf
[91] Recommended Practice API 1540 – Design, Construction, Operation and Maintenance of Aviation Fuelling Facilities, 4th Ed. (2004)

[92] A Sensitivity Analysis of a Computer Model-Based Leak Detection System for Oil Pipelines – Lu, She and Loewen – Energies MDPI
[93] State Estimation of Pipeline Models using the Ensemble Kalman Filter, Jason Modisette, PSIG 2013
[94] Application Benefit of the Online Simulation Software for Gassco's Subsea Pipeline Network, Velde, Postvoll (Gassco), Hanmer, Munro (Atmos Int.)
[95] Presentation at Pipeline Simulation Interest Group (PSIG) 2013.
[96] How Pipelines Make the Oil Market Work – Their Networks, Operation and Regulation, Allegro
[97] Modelling & Simulation of Gassco Wet Gas Pipelines PSIG 1604, Hanmer, Basnett, Issak, Rinde.
[98] Modelling transient leaks from pressure vessels including effects of safety systems, Stene, Harper and Witlox – IChemE Symposium 159, Hazards 24
[99] antoine.frostburg.edu/chem/senese/101/liquids/faq/antoine-vapor-pressure.shtml
[100] Improving Operational Safety in Oil Pipelines, Schraml et. al., Pipeline Technology Conference 2011
[101] A Slack Flow Model With Moving Regime Boundaries, Jason Modisette, PSIG 2018
[102] https://www.iea.org/reports/the-future-of-hydrogen - 2019
[103] thechemicalengineer.com/features/clean-hydrogen-part-1-hydrogen-from-natural-gas-through-cost-effective-co2-capture/
[104] thechemicalengineer.com/features/hynet-demonstrating-an-integrated-hydrogen-economy/
[105] decarbconnect.com/h21-project-testing-the-route-to-domestic-and-industrial-hydrogen-distribution/
[106] www.spiraxsarco.com/learn-about-steam/steam-distribution/pipes-and-pipe-sizing
[107] Technical assessment of Hydrogen transport, compression, processing offshore, North Sea Energy
[108] Transporting Hydrogen-Natural Gas Mixtures, Rainer Kurz et. al., Solar Turbines, PSIG 2110 (2021)
[109] turbomachinerymag.com/view/readying-pipeline-compressor-stations-for-100-hydrogen
[110] The Use Of The Natural-Gas Pipeline Infrastructure For Hydrogen Transport In A Changing Market Structure, Dries Haeseldonck, William D'haeseleer, K.U. Leuven
[111] An integrated solution for planning and operating Power-To-Gas facilities in coupled gas and electricity networks, Adolfo et. al., PSIG 2115 (2021).
[112] Pipeline Transient Optimization for a Gas-Electric Coordination Decision Support System, Zlotnik et. al., PSIG 1919 (2019).
[113] Multivariable analysis for selection of natural gas compression technology considering the impact of Solar PV power generation, Alvarez et. al. PSIG 1917 (2019)
[114] https://afdc.energy.gov/files/pdfs/hyd_economy_bossel_eliasson.pdf
[115] Hazardous Liquid Pipeline Rupture and Release Marshall, Michigan July 25, 2010 – Accident Report ntsb.gov/investigations/AccidentReports/Reports/PAR1201.pdf